INFORMATICA

INFORMATICA

Mastering Information through
the Ages

Alex Wright

CORNELL UNIVERSITY PRESS   ITHACA AND LONDON

Hardcover edition originally published as *Glut: Mastering Information through the Ages* 2007 by Joseph Henry Press

Glut paperback edition published 2008 by Cornell University Press with permission, all rights reserved
Glut first printing, Cornell Paperbacks, 2008

Informatica first published 2023 by Cornell University Press

Librarians: A CIP catalog record for this book is available from the Library of Congress.

ISBN 978-1-5017-6867-5 (paperback)
ISBN 978-1-5017-6868-2 (pdf)
ISBN 978-1-5017-6869-9 (epub)

For my parents

Our own Middle Age, it has been said, will be an age of "permanent transition," for which new methods of adjustment will have to be employed . . . an immense work of bricolage, balanced among hope, nostalgia and despair.
—Umberto Eco, "Living in the New Middle Ages"

Contents

When *Glut* first appeared in early 2007, the Internet was a simpler place. Most of the 1.2 billion people who interacted with the global network that year did so using a web browser on a desktop or laptop computer. Popular websites of the day included Yahoo!, Myspace, and AOL—long-faded relics of that bygone digital era. Twitter had launched just a few months earlier. YouTube had not yet turned two. A team of designers at Apple was still working behind closed doors on the first version of the iPhone, which would debut later that year.

As I write this today in January 2022, the Internet has blossomed into a far richer, more complex, and polymorphous place. The number of users has soared to more than four billion, most of them now accessing the network using smartphones with a constellation of special-purpose apps. The list of most trafficked Internet services ("websites" seems like a quaintly dated term) now includes the likes of TikTok, Twitch, WhatsApp, and other apps through which a seemingly endless gusher of words, pictures, sounds, and moving images flashes briefly into view. The years ahead promise even more dramatic changes to the global information ecosystem. Artificial intelligence systems can now intuit our needs and preferences with increasing precision, provide recommendations, and even generate new forms of knowledge in the shape of computer-generated news stories, visual imagery, and even poetry. Emerging web3 platforms built around digital blockchains augur other new forms of information (like nonfungible tokens). Facebook, now known as Meta, is investing billions of dollars into creating augmented and virtual reality experiences that may further reshape the experience of creating, collecting, and consuming information.

The warp and weft of human knowledge continues to evolve and with it the interconnected social, cultural, political, and economic systems that are the lifeblood of human civilization. While *Glut* focused intentionally on surveying the history of information systems that preceded the World Wide Web—rather than trying to tell the still-evolving story of a new technology—all history is inevitably written from the vantage point of the present day. Were I writing *Glut* from scratch today, it would almost certainly take on a different form. In the horse-and-buggy era of the 2007 web, the promise of the Internet seemed to carry echoes of the long-held human dream of the universal library. As such, *Glut* can best be understood as a kind of prehistory of the web. Today, increasingly it seems that the web is designed to become just one of many platforms for interacting with an endlessly evolving digital environment, many of them hidden from public view in so-called walled gardens. Trying to tell the history of all these emerging forms of expression—social media, cryptocurrency, algorithmic natural language processing, to name but a few—would demand entirely new books. Fortunately, many of those books have since been written. Why, then, revisit this one?

By 2007, the explosive growth of the Internet had already brought into stark relief the problem of organizing large bodies of recorded information for public consumption. But there was still an air of excitement and utopian zeal surrounding the promise of the global network. That rhetorical optimism has now largely faded from view. Today, the unintended consequences of unchecked technotopianism have come more clearly into view. Over the past fifteen years, we have come to understand the deleterious societal effects of neoliberal market dynamics surrounding user-generated content, the extractive nature of surveillance capitalism on a global scale, and the rise of a dangerous factionalism in many corners of the world, fueled by disinformation flowing through social media networks. The Internet now touches, and has reshaped, nearly every aspect of contemporary life, in ways both good and bad. Against a backdrop of convulsive social, political, and economic transformation, the problems of information management may seem esoteric by comparison. But the central challenge that *Glut* explored remains: humanity is creating an ever-increasing outpouring of recorded knowledge, and our systems for collecting, managing, and making sense of that knowledge are becoming increasingly frayed. But, as I try to show in *Informatica*, this problem is hardly a new one. By exploring our present dilemmas through a historical lens, my hope is to locate avenues for further exploration, and promising ideas left by the historical wayside that might yet help us envision a more humane, ethical, and sustainable information ecosystem.

In its earliest incarnation, the World Wide Web seemed to hold the potential of evolving into the kind of universal library that features in many of the utopian visions mentioned in *Glut*—the royal libraries of Ptolemy, Shi Huangdi, and Charlemagne, or latter-day visions like Jorge Luis Borges's Library of Babel, the Mundaneum, Xanadu, and the Knowledge Navigator, to name a few. *Glut's* central argument—that the dream of universal knowledge has recurred regularly at key technological inflection points throughout human history—seems borne out by the events of the past three decades, ever since Tim Berners-Lee released his first open-source web browser.

Much of the unchecked optimism that accompanied the first flowering of the Internet in the mid-1990s has since given way to more critical and dystopian narrative about big tech monopolies, surveillance capitalism, the ethics of artificial intelligence, and data privacy. Yet, the central premise of *Glut* persists: that we all share a common impulse toward gathering, organizing, and distributing information—to create a collective knowledge edifice that can sustain the culture at large. But this is also the first era in which great fortunes have been built on such an enterprise. The efforts of past eras at creating universal knowledge reservoirs typically happened under the auspices of empires, nation-states, universities, and foundations. Only in the past thirty years has this ancient endeavor become the province of for-profit corporations. The influence of industrial capitalism and the consumer economy has shaped and distorted the information landscape in countless ways, bending this time-honored pursuit toward the service of commercial ends. Yet at the same time, the innovative impulse at the heart of the tech industry has unleashed an unprecedented wave of creative self-expression, as new forms of communication and meaning making, especially in the realm of social media, have captured humanity's imagination at a pace that would have bewildered the sober archivists of generations past.

Technology skeptics might take heart in the knowledge that if there seems to be one reliable pattern in humanity's efforts to collect and organize the world's knowledge, it is that these efforts rarely succeed over the long term. Things change. New forms of expression arise, and with them new strategies and mechanisms for leveraging humanity's shared experiences. As organizational structures congeal into hierarchies, new networks come along to disrupt them. The forms and structures of human knowledge will undoubtedly continue to evolve, but even as particular forms may come and go, broader patterns may prevail: the tension between hierarchical and networked systems, the inherent impermanence of all recorded knowledge, and the cascade of unanticipated consequences that often accompany the introduction of new information technologies at scale.

Glut took as its premise the assumption that organizing and managing information would continue to pose a struggle as the world's collective data stores became more networked. And although that premise seems durable enough, two major trends have emerged since the book's publication that seem to demand some kind of reckoning here. The first, as mentioned above, is the rise of mobile computing. The increasing availability of geo-coded data, accelerometers that can measure and report the location and orientation of a device with precision within a few millimeters, and the emerging enabling technologies of augmented and virtual reality point toward an environment where data begin to weave into the three-dimensional world around us, suggesting that we might do well to investigate the heritage of architecture and way-finding systems in looking for historical reference points into these kinds of experiential interfaces. The second is the rise of social media and the vast outpouring of human expression into the sharing of experiences and conversations atop these platforms.

Although social media had emerged into its infancy when *Glut* was released, it seems fair to say that none of the early progenitors of hypertext discussed in those pages came anywhere close to anticipating the emergence of a global networked conversation involving billions of human beings, let alone the new forms of expression that would take shape across these new platforms. In retrospect, Walter J. Ong's work on secondary orality comes closest to anticipating the rise of social media (see chapter 5). In hindsight, Ong's vision seems more central than I could have anticipated. If I were starting from scratch today, I might ground this inquiry in a deeper exploration of the tensions between oral and literate forms of discourse—a dynamic we can see playing out regularly today in the uneasy relationship between the traditional institutional keepers of the literate tradition (e.g., newspapers and book publishers) and the emergent upstart social media platforms that traffic by and large in the kind of secondary orality that Ong predicted. Fortunately, recent years have seen capable authors like Tom Standage explore the historical antecedents to social media, while others have probed deeply into the history of classification, notably Markus Krajewski, whose work on the history of cataloging cards has opened up new dimensions of inquiry into the history of classification, and Colin Burke, whose work on Herbert Haviland Field marks a signal contribution to the history of information science.[1]

In the years since *Glut* was released, I have also had the chance to collect feedback from a number of engaged readers, as well as a few constructive critics, who have pointed out blind spots in this historical narrative: especially involving other early progenitors of networked information systems whom I managed to overlook, including seminal figures like Suzanne Briet, Watson Davis, Wendy Hall, Conrad Gessner, Herbert Field, and especially Claude Shannon—whose foundational work on information theory underlies much of the subsequent development of the technology industry in the twentieth century. And although much of the book tried to embrace a wide-angle view of information systems across a range of cultures, the later chapters dealing with the nineteenth- and twentieth-century history of information systems centered on a preponderance of white men. While their contributions matter greatly to the subsequent development of the Internet, it is worth acknowledging the risks of succumbing to the European great white man view of history, bound up as it is with the relentless techno-optimism and underlying systematic oppression that have come into such stark focus over recent years.

In the revisions to these chapters, I have tried to broaden the cultural perspectives to include the important contributions of Islamic scholars to the preservation of classical knowledge during the so-called Dark Ages in Europe and of Chinese and Korean printers to the development of movable type in the centuries preceding Gutenberg. I have also endeavored to strike a balance between continuing to acknowledge the important contributions of underappreciated figures like Paul Otlet, J. C. R. Licklider, and Douglas Engelbart while also acknowledging the systems of colonial oppression and cultural imperialism within which their work took shape.

In the years since *Glut*'s publication, I spent considerable time deepening my research into Otlet and his contemporaries in the European documentalist movement of the 1920s and 1930s. This work culminated in my 2014 book on Otlet, *Cataloging the World*.[2] Building on that body of research, I have expanded the section on Otlet and his circle in chapter 10. I have also addressed other assorted errors of omission and commission along the way, and would like to express my gratitude to those readers who took the time to share their feedback and point out opportunities for strengthening some of the lines of arguments presented herein. As ever, whatever mistakes remain are mine alone.

INTRODUCTION

Ever since the Internet emerged into the public consciousness at the end of the twentieth century, we have seen a bull market in hyperbole about the digital age. Visiting San Francisco at the height of the 1990s dot-com boom, Tom Wolfe noted the particular brand of euphoria then sweeping the city. Wolfe, who made his journalistic bones chronicling the psychedelic raptures of the city's 1960s pranksters, spotted a similar strain of quasi-mystical fervor taking hold among the young acolytes of the digital revolution.[1] "Much of the sublime lift came from something loftier than overnight IPO billions," he wrote, "something verging on the spiritual." Enthusiastic dot-commers "were doing more than simply developing computers and creating a new wonder medium, the Internet. Far more. The Force was with them. They were spinning a seamless web over all the earth."[2] In the Day-Glo pages of *Wired* and a host of also-ran new economy magazines, the so-called digerati were pumping a rhetorical bubble no less inflated than the era's IPO-fueled stock prices. The writer Steven Johnson compared the dawning age of software to a religious awakening, predicting that "the visual metaphors of interface design will eventually acquire a richness and profundity that rival those of Hinduism or Christianity."[3] Elsewhere, the supercomputer pioneer Danny Hillis argued that the advent of the World Wide Web signaled an evolutionary event on par with the emergence of a new species: "We're taking off," he wrote. "We are not evolution's ultimate product. There's something coming after us, and I imagine it is something wonderful. But we may never be able to comprehend it, any more than a caterpillar can imagine turning into a butterfly."[4] More recently, the inventor and futurist Ray Kurzweil has gone so far as to suggest that we are undergoing

1

a "technological change so rapid and profound it represents a rupture in the fabric of human history," an event so momentous that it will trigger "the merger of biological and nonbiological intelligence, immortal software-based humans, and ultra-high levels of intelligence that expand outward in the universe at the speed of light."[5] Could the arhats themselves have painted a more dazzling picture of enlightenment?

Mystical beliefs about technology are nothing new, of course. In 1938, H. G. Wells predicted that "the whole human memory can be, and probably in a short time will be, made accessible to every individual," forming a so-called world brain that would eventually give birth to a "widespread world intelligence conscious of itself."[6] Similar visions of an emerging planetary intelligence surfaced in the mid-twentieth century writings of the Catholic mystic Pierre Teilhard de Chardin, who foresaw the rise of an "extraordinary network of radio and television communication which already links us all in a sort of 'etherised' human consciousness." He also anticipated the significance of "those astonishing electronic computers which enhance the 'speed of thought' and pave the way for a revolution," a development he felt sure would spur the development of a new "nervous system for humanity" that would ultimately coalesce into "a single, organized, unbroken membrane over the earth."[7] Teilhard believed that this burgeoning networked consciousness signaled a new stage in God's evolutionary plan in which human beings would coalesce into a new kind of social organism, complete with a nervous system and brain that would eventually spring to life of its own accord. Teilhard never published his writings during his lifetime—the Catholic Church forbade him from doing so—but his essays found an enthusiastic cult following among fellow Catholics like Marshall McLuhan, who took Teilhard's vision as a starting point for formulating his theory of the global village.

Today, the torch song of technological transcendentalism has passed from the visionary fringe into the cultural mainstream. Scarcely a day goes by without some hopeful dispatch about new web applications, digital libraries, or munificent technocapitalists spending billions to wire the developing world. Some apostles of digitization argue that the expanding global network will do more than just improve people's lives; it will change the shape of human knowledge itself. Digital texts will supplant physical ones, books will mingle with blogs, and fusty old library catalogs will give way to the liberating pixie dust of Google searches. As the network sets information free from old physical shackles, people the world over will join in a technological great awakening.

Amid this gusher of cyberoptimism, a few dissidents have questioned the dark side of digitization: our fracturing attention spans, the threats to personal privacy, and the risks of creeping groupthink in a relentlessly networked world. "We may even now be in the first stages of a process of social collectivization that will over time all but vanquish the ideal of the isolated individual," writes the critic Sven Birkerts. In this dystopian view, the rise of digital media marks an era of information overload in which our shared cultural reference points will dissolve into a rising tide of digital cruft.[8]

For all the barrels of ink and billions of pixels spent chronicling the rise of the Internet in recent years, surprisingly few writers seem disposed to look in any direction but forward. "Computer theory is currently so successful," writes the philosopher-programmer Werner Künzel, "that it has no use for its own history."[9] This relentless fixation on the future may have something to do with the inherent "forwardness" of computers, powered as they are by the logics of linear progression and lateral sequencing. The computer creates a teleology of forward progress that, as Birkerts puts it, "works against historical perception."[10]

In times past, when people felt their lives affected by new information technologies—like symbols, alphabetic writing, or the printing press—they looked for ordering principles to help them make sense of a changing world. They invented mythologies, cosmic hierarchies, library catalogs, encyclopedias, and so on. Whatever strengths and shortcomings these systems may have had, they all shared one essential trait: transparency. The logics of Aristotle or the library catalog are plainly visible to anyone who cares to look into them. Today, however, we put our faith in mechanisms we cannot see and that few of us will ever understand: the secret algorithms of Google, Amazon's recommendation engine, or fuzzy fabrications like "collective intelligence." As we entrust more and more of what we know to these increasingly opaque systems, we are growing increasingly reliant on an elite priesthood of private-sector programmers, ministering behind closed doors to the Oracles in the server room. When people feel their lives affected by forces they do not understand, they may start to imagine the presence of supernatural forces at work. They may see ghosts in the machine.

My aim in writing this book is to resist the tug of mystical technofuturism and approach the story of the information age by looking squarely backward. This is a story we are only beginning to understand. Like the narrator in Edward Abbott's *Flatland*—a two-dimensional creature who wakes up one day to find himself living in a three-dimensional world—we are just starting to recognize the contours of a broader information ecology that has always surrounded us. Just as human beings had no concept of oral culture until they learned how to write, so the arrival of digital culture has given us a reference point for understanding the analog age. As McLuhan put it, "One thing about which fish are completely unaware is the water, since they have no *anti-environment* that would allow them to perceive the element they swim in." From the vantage point of the digital age, we can approach the history of the information age in a new light.

To do so requires stepping outside of traditional disciplinary constructs, however, in search of a new storyline.

In these pages, I traverse a number of topics not usually brought together in one volume: evolutionary biology, cultural anthropology, mythology, monasticism, the history of printing, the scientific method, eighteenth-century taxonomies, Victorian librarianship, and the early history of computers, to name a few. No writer could ever hope to master all of these subjects. I am indebted to the many scholars whose work I have relied on in the course of researching this book. Whatever truth this book contains belongs to them; the mistakes are mine alone.

I am keenly aware of the possible objections to a book like this one. Academic historians tend to look askance at "meta"-histories that go in search of long-term cultural trajectories. This is a synthetic work that covers a lot of historical ground, and in some cases I have knowingly committed the sin of citing secondary sources where, as an independent scholar without a university affiliation, I was unable to gain access to primary source material. As a generalist, I knowingly run the risks of intellectual hubris, caprice, and dilettantism. But I have done my homework, and I expect to be judged by scholarly standards. This work is, nonetheless, fated to incompleteness. Like an ancient cartographer trying to draw a map of distant lands, I have probably made errors of omission and commission; I may have missed whole continents. But even the most egregious mistakes have their place in the process of discovery. And perhaps I can take a little solace in knowing that Carl Linnaeus, the father of modern biology, was a devout believer in unicorns.

NETWORKS AND HIERARCHIES

I am a firm believer that without speculation there is no good and original observation.

—Charles Darwin, letter to A. R. Wallace, 1857

When the Spanish conquistadores first encountered the Zuni people of the North American Southwest, they noticed something strange about their villages. The tribe had divided each of its six pueblos (as the Spanish called them) into a set of identical quadrants, aligned with the four points of the compass. Each quadrant housed a troop of clans within the larger tribe: the clans of the Crane, the Grouse, and the Evergreen lived in the north; the clans of Tobacco, Maize, and Badgers lived in the south. Each clan enjoyed a set of special relationships with the natural world. To the people of the north belonged wind, winter, and the color yellow. The people of the west knew water, spring, and the color blue. The people of the north made war. The people of the west kept the peace. When the villagers sat together, they sat apart, like hawks and doves. To the four cardinal directions, the Zuni added three vertical ones: the sky, the earth, and a middle realm in between. In the sky, all the colors of the world swirled together; down below, the earthen realm was black. In the middle realm, everything came together; heaven and earth were joined. To each of these seven directions, the Zuni assigned everything in the cosmos: animals, natural elements, supernatural forces, social responsibilities, families, and individual members of the tribe. This all-encompassing system equipped the Zunis with a taxonomy of the natural world, a social and political system, a mythology, and a framework for spiritual belief.[1]

The Zuni system represents one people's solution to a problem we all share: how to manage our collective intellectual capital. For more than a hundred thousand years, human beings have been collecting, organizing, and sharing information, creating systems as varied as the cultures that produced them. Along the way, they have invented a panoply of semantic tools: taxonomies, mythologies, temple archives, books, libraries, indexes, encyclopedias, and in recent years, digital computers.

Today, we live in an age of exploding access to information, awash in what Richard Saul Wurman calls a tsunami of data.[2] In 2006 (when *Glut* was written), human beings produced more than five exabytes' worth of recorded information per year:[3] documents, e-mail messages, television shows, radio broadcasts, web pages, medical records, spreadsheets, presentations, and books like this one. That is more than fifty thousand times the number of words stored in the Library of Congress, or more than the total number of words ever spoken by human beings.[4] Since then, the volume of global data production has continued to accelerate drastically. From 2010 to 2020, the world's data stores expanded fiftyfold, to an estimated forty thousand exabytes of data in 2020.[5] By 2025, that number may rise as high as 175,000 exabytes.[6] Amid this welter of bits, perhaps some of us worry, like Plato's King Thamus, whether our dependence on the written record will weaken our characters and create forgetfulness in our souls.

As the proliferation of digital media accelerates, many of us are witnessing profound social, cultural, and political transformations whose long-term outcome we cannot begin to foresee. Organizational charts are flattening, as electronic communication tools enable employees to bypass old chains of command; national borders are growing more porous, as networked data flow across old boundaries; and long-established institutional knowledge systems (e.g., library catalogs) are fast becoming anachronisms in the age of web search engines. Wherever networked systems take root, it seems, they disrupt the old hierarchical systems that preceded them. Indeed, a faith in the death of hierarchy has become one of the most durable nostrums of the digital age. In the popular 1999 tract *The Cluetrain Manifesto*, the authors proposed a credo for the Internet age: "hyperlinks subvert hierarchy."[7] That sentiment captures the widely held belief that the rise of the Internet signals the permanent disruption of old institutional bureaucracies and the birth of a new enlightened age of individual expression: a new renaissance of creativity and personal freedom. In this utopian view, hierarchical systems are restrictive oppressive tools of control, while networks are open democratic vehicles of personal liberation. When networks triumph over hierarchies, then, humanity takes a great leap forward. Manuel Castells goes so far as to say that the networked revolution "represents a qualitative change in the human experience."[8]

This comforting narrative is too tidy by half. Networked information systems are by no means entirely modern phenomena, nor are hierarchical systems necessarily doomed to extinction. There is a deeper story at work here. The fundamental tension between networks and hierarchies has been percolating for eons. Today, we are witnessing the latest installment in a long evolutionary drama.

Since the words *network* and *hierarchy* will recur throughout this book, let me spend a moment with the terms. A hierarchy is a system of nested groups. For example, an organization chart is a kind of hierarchy in which employees are grouped into departments, which in turn are grouped into higher-level organizational units, and so on, up to the top rung of the management ladder. Other kinds of hierarchies include government bureaucracies, biological taxonomies, or a system of menus in a software application. The computer scientist Jeff Hawkins suggests that human memory itself can be explained as a system of nested hierarchies running atop a neural network.[9] A network, by contrast, emerges from the bottom up; individuals function as autonomous nodes, negotiating their own relationships, forging ties, coalescing into clusters. There is no "top" in a network; each node is equal and self-directed. Pure democracy is a kind of network; so is a flock of birds or the World Wide Web.

Networks recur throughout the natural world. From the mitochondrial networks of simple cells to the circulatory systems of animals, from the neural networks of the brain to the complex interactions of social organisms like termites, ants, chimpanzees, and people—the topology of networks shapes the world around us. Indeed, sociologists Nicholas Christakis and James Fowler have even suggested that our species should be known as *Homo dictyous*, "network man."[10]

Networks and hierarchies are not mutually exclusive, however; indeed, they usually coexist. The historian Niall Ferguson even goes so far as to suggest that a hierarchy is, in essence, just a particular form of network, one with a singular node at the top.[11] We might, for example, work for a company with a formal organization chart; at the same time, we probably also maintain a personal network of colleagues that has no explicit representation in the formal organization: a network within a hierarchy. Similarly, the Internet, ostensibly a pure network, is actually composed of numerous smaller hierarchical systems.

At its technical core, the Internet works by breaking large collections of data into small packets, tiny hierarchical units of information stored on a server, which are then dispersed across the network and reassembled in a client application such as a web browser. At a higher level, much of the content of the web is generated within organizational hierarchies—like companies, educational and nonprofit institutions, and government agencies—as well as by ostensibly self-directed individuals who nonetheless rely on hierarchical organizations (e.g., computer manufacturers or service providers) to participate in the network. And for all the seeming flatness of the global Internet, most of us make sense of the web in hierarchical terms: by navigating through menus on a website, for example, or selecting

FIGURE 1. Hierarchy © Alex Wright, 2022.

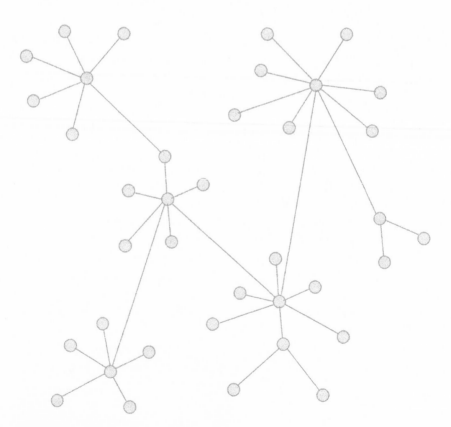

FIGURE 2. Network © Alex Wright, 2022.

from a narrow list of search results. In other words, networks and hierarchies not only coexist, but they continually give rise to each other.

Science writer Howard Bloom has suggested that the tension between networks and hierarchies is not an exclusively human phenomenon but part of a deeper process embedded in the fabric of the universe itself, stretching all the way back to the big bang.[12] We need not look quite so far back, however. Two billion years will do.

Early Information Systems

Although our reliance on networked information systems may seem like an acutely modern dilemma, we are not the first generation—or even the first species—to wrestle with the problem of information overload. In fact, the information age started not with microchips or movable type but with the first flowering of complex life. To approach the history of information systems from a purely human-centered perspective is to overlook the lessons of billions of years' worth of evolutionary history. Just as our brains carry around some very old reptilian equipment, so our collective strategies for managing data bear the traces of patterns that took shape long ago.

John Locke famously argued that "beasts abstract not." But in recent years, thanks to the breakthrough work of geneticists and evolutionary psychologists, we are beginning to appreciate the surprising complexity of nature's nonhuman information systems. The animal world is rife with examples of organisms developing collective strategies for accumulating, storing, and distributing information in groups.

If we accept the familiar construct of information as lying on a continuum from data to wisdom (data > information > knowledge > wisdom), then there is no question that other animals create, share, and organize information. And although animal minds may not process the abstractions of human thought, they frequently employ information-pooling strategies that bear startling resemblances to our own.

In 1970, biologist Lynn Margulis proposed a revolutionary hypothesis about the origin of complex organisms, suggesting that the relationship between networked and hierarchical systems is deeply woven into the fabric of life itself. To make a particularly long story short, about two billion years ago, the first multicellular organisms, eukaryotes, took shape as host bacteria began to allow other bacteria to take up residence within them, gradually conscripting their formerly independent siblings into a kind of cellular serfdom. These early complex life-forms came into existence not as fully formed organisms but rather through a gradual process of evolving social cooperation. Eventually, as teams of bacteria began to reproduce in tandem with their host organisms, they coalesced into nucleated cells capable of reproducing themselves. The first multicellular life-forms came into being, then, as self-organizing networks. Gradually, those networks gave rise to an emergent hierarchy: the complex organism. These self-directed biological hierarchies in turn begin to interact with one another, forming new higher-level networks that in turn coalesced into more complex biological hierarchies. From this escalating fugue of hierarchies and networks, life has evolved.

By 500 million BCE, complex organisms of a trillion cells or more began to take shape: the first insects, clams, and birds, equipped with highly specialized organs like eyes, nerve cells, and memory coils. As their neural powers grew, these animals became more autonomous. But independence came at a price: the more complex life-forms became, the more isolated they became from one another. To compensate for the loss of old physical bonds, they began to develop what maverick scientist Howard Bloom calls the synapse of the social brain: imitation.[13] By following each other's behavioral cues, certain animals could pool their sensory data, processing collective experience with a capacity far exceeding the abilities of a lone organism. Imitation provided the glue that bound these now-autonomous animals into a higher order of organization: social networks. They became, in effect, superorganisms.

A biological superorganism—like an insect colony, a flock of birds, a school of fish, or arguably, certain groups of humans—is both a network and a hierarchy. It emerges from the networked interaction of individual organisms, in turn giving rise to higher-order hierarchies. As individual organisms transmit information to each other, they strengthen the bonds that unite the group. But what exactly is being transmitted? Social information is after all noncorporeal; it is not a physical "thing" (even though it may take expression in the physical environment). Yet, there is no question that animals are transmitting some kind of "thing" to each other. What is it?

In his influential 1975 book *The Selfish Gene*, evolutionary biologist Richard Dawkins dubbed that thing a meme.[14] Other biologists have variously labeled it a mnemotype, idea, idene, sociogene, or concept. But only meme has managed to penetrate the popular vernacular. Whatever term we use, there is no question that other animals regularly record, share, and preserve information that has no physical manifestation. Biologists have learned a great deal about how memes travel between members of a social group, mapping the trajectories of information exchange from individual to individual; less well understood are the higher-level dynamics that characterize the interactions of memes at the group level.

One highly speculative explanation of how nature's "mass minds" operate comes from Howard Bloom, whose unconventional theories and colorful rhetoric have marked him as an *agent provocateur* in mainstream scientific circles. Bloom postulates that the phenomenon of collective intelligence emerges from the interplay of five essential forces: conformity enforcers, diversity generators, inner judges, resource shifters, and intergroup tournaments. Conformity enforcers (e.g., worker bees or middle managers) ensure that the group as a whole maintains sufficient cohesion to survive adverse conditions. Diversity generators (e.g., stray ants or artists) are the "odd ducks" who generate alternative hypotheses for the group to consider, thus ensuring variation. Intergroup tournaments (e.g., the waggle dances of bees or scientific debates) enable societies to test alternative hypotheses. Inner judges reward productive behavior and punish deleterious actions. Finally, resource shifters (e.g., alpha chimpanzees or corporate executives) make sure successful adaptations receive the support they need to benefit the group as a whole.[15] Bloom's model, though intriguing, is too figurative to pass any empirical test. Nonetheless, respected evolutionary biologists like Lynn Margulis and David Sloan Wilson have recognized value in Bloom's original, if decidedly left-field, conception of the global brain. We do not have to accept Bloom's theory as hard science, however, to appreciate it as metaphor. As Alfred North Whitehead put it, "It is more important that a proposition be interesting than that it be true. But of course a true theory is more apt to be interesting than a false one."

Today, we can find ample evidence throughout the natural world of networked superorganisms pooling information, organizing and distributing that information to the right individuals at the right time, and preserving successful group strategies for future generations. Insect societies provide the most often cited example of how seemingly simple creatures can process information with a sophistication seemingly not predicted by their genetic blueprints. In 1946, biologist Karl von Frisch won a Nobel Prize for deciphering the syntax of bees' famous waggle dances, the carefully choreographed tournaments that enable beehives to identify the location of food supplies with astonishing precision. Certain bees scout a particular area for potential food sources and then return to the hive to present choreographed "reports" on what they found. As each bee makes its presentation, the group as a whole registers its level of enthusiasm and compares incoming reports to make a group decision about where to look for food next. No individual bee possesses anywhere near the smarts to make such a decision, but as a group, the bees generate a collective "mind" far more clever than the sum of its tiny-brained parts. Elsewhere, studies have shown that although an individual honeybee can retain a piece of data in its memory bank for up to six days, the hive as a whole can retain that same piece of data for up to three months—double the life span of a single bee.[16]

Some insect colonies display even higher-order behaviors that rely not just on social displays but that actually involve coding their memes onto the physical environment. In 1959, biologist Pierre-Paul Grassé coined the term *stigmergy* (meaning "incite to work") after studying how termites erect their monumental nests.[17] Grassé observed that termites build nests by dint of a brutally simple algorithm. A single termite carries a chewed pellet of dirt in its jaws; whenever it encounters a slightly elevated mound of dirt, it drops the pellet. When a number of termites begin to drop pellets on the same mound, a small clump begins to form; other passing termites, detecting the clump, begin to drop their own pellets, and the mound grows. If one mound starts growing in close proximity to another, the termites will begin scuttling back and forth between the mounds, building diagonally to join them together. From these simple rules, great cathedral-like artifacts emerge. They are living information systems.

Ants, too, exhibit a capacity for stigmergy, communicating to each other through pheromone trails that guide their collective behavior. In one study, researcher Jean-Louis Deneubourg provided an ant colony with two separate tree branches, each a pathway to the same food supply. Individual ants immediately began charging blindly down both paths. Within a few minutes, however, the entire colony had collectively chosen the shorter of the two branches. As the first wave of ants returned from their round-trip journey, they began signaling their remaining compatriots to follow their trail. As the collective pheromone trail grew stronger, the entire colony abandoned the unscented trail and flocked down the fast-food path.

Ants' dual roles as both actors and signals make them a compelling reference model for computer programmers. The new field of swarm intelligence has grown up in recent years to explore the application of insect-like solutions to practical problems like computer network design. At Hewlett-Packard, a research team developed a network routing system based explicitly on the tracking behaviors of ant colonies, consisting of a large number of tiny computer programs that wound their way through a distributed network, searching out the least congested routes. Depositing "virtual pheromone" trails at each node, the swarm of programs quickly deduced the fastest path through the network.

Whereas other forms of communication in the natural world typically involve one-to-one transmission between individuals, stigmergy involves at least one step of intermediation: encoding information on an external object. It is a mechanism for indirect collaboration, equipping life-forms with what philosopher Ron McClamrock describes as the ability to "enforce on the environment certain kinds of stable properties that will lessen our computational burdens."[18] In other words, stigmergy allows social groups to harness the physical world as a memetic storehouse, something at which human beings have long excelled.

Whereas insect societies provide nature's most elegant examples of information systems that transcend the capacities of individuals, the animal kingdom is rife with other examples of species that share information in groups. In an oft-cited bird study in 1950s England, a small population of English tits stumbled upon the secret of pecking through aluminum milk bottle caps as a means to get at the milk inside. The new strategy soon took hold among the tit population at large, and an avian meme took flight. An epidemic of bottle pecking spread across the surrounding counties, as successive generations of tits learned the fine art of milk larceny from their peers. In New Zealand, flocks of oystercatcher birds have developed their own idiosyncratic customs for cracking open mussel shells: some groups hammer the shell until it opens; others pry it open with their beaks. All oystercatchers are born with the capacity for either behavior; the sole determinant of how an individual bird behaves appears to rest entirely on the bird's social affiliation. In other words, social networks, not genetics, determine the persistence of certain behaviors from generation to generation.

Whereas insects and birds demonstrate how information systems emerge in species whose intelligence ranks far below our own, we can find examples of complex social networking behaviors in more intelligent animals, notably our cousins the primates. In Japan, a troop of hungry macaques stumbled upon an innovative process for separating edible wheat from indigestible sand. On one island, an enterprising macaque came up with a handy trick: by throwing the entire mixture of wheat and sand into the sea, the macaque discovered that the wheat would float to the top while the sand sank to the bottom. Success begat imitation; soon after the first macaque stumbled upon the wheat-sifting technique, the practice spread through imitative "word of mouth"; within five years, the entire macaque population of Japan had learned to sift their wheat.[19]

Our closest cousins, chimpanzees, also display highly localized traditions that persist across generations. Some chimpanzee troops use rocks to break open nuts; others rely on tree trunks. Some groups have figured out how to use twigs as rudimentary "fishing rods" for digging up ants; others forge long hooked branches into tools for retrieving figs. These local traditions get passed down through generations purely through social transmission, not genetic inheritance. In a similar study, biological anthropologist Carel van Schaik and his colleagues discovered a group of orangutans on the western coast of Sumatra that had developed highly localized customs for using tools. One group had perfected an intricate procedure for removing seeds from the toughened husks of the *Neesia* tree by stripping the bark from twigs and using the rudimentary tools to jimmy open the husks.[20] This particular trick seems to have emerged only within one local population; other nearby groups in similar environments had no such innovation, and their *Neesia* seeds went uneaten. The original group had also developed a series of other tool-use innovations, far outstripping the achievements of the surrounding communities. Why did one group of orangutans prove so consistently innovative while their nearby cousins languished? The key to cultural innovation, it seems, has something to do with population density. Van Schaik hypothesizes that "populations in which individuals had more chances to observe others in action would show a greater diversity of learned skills than would populations offering fewer learning opportunities. And indeed, we were able to confirm that sites in which individuals spend more time with others have greater repertoires of learned innovations."[21] The orangutans of Saaq crossed a threshold of social proximity that enabled them to adapt more quickly and successfully to their environment. As we will see later in the book, a similar phenomenon recurs among human social groups, where population density appears to determine the velocity of technological change. Social concentration, it seems, is the engine of innovation.

In addition to imitative learning, primates traffic in another more subtle form of information exchange: emotion. We may think of our own emotions in highly personal terms such as joy, sadness, anger, jealousy, and love. Across the primate kingdom, however, emotional expression plays another broader social role in facilitating the transfer of memes between group members. Recent primate research also suggests that emotional expression may have something to do with our capacity for symbolic expression. "Symbol formation results from a series of stages of affective transformations," writes psychiatrist Stanley Greenspan, suggesting that symbols arose not from the use of human spoken language but from a deeper emotional wellspring grounded in the primate limbic system (the part of the brain that controls emotions). Greenspan has formulated an intriguing theory that suggests an evolutionary progression of symbolic language: from engagement and signaling (e.g., reacting to a mother's facial expressions) to simple call-and-response interactions (e.g., sharing environmental data) to elaborate greeting rituals (e.g., courtship or coalition building) to "co-regulated affective signaling" (e.g., organizing hunting parties with differentiated roles).[22] Greenspan believes that this progression of signaling behavior evolved in concert with a series of genetic mutations among our primate ancestors that would ultimately equip *Homo sapiens* with our uniquely developed capacity for symbolic expression.

If this startling theory of human symbolism holds true, it suggests a radical rethinking of the origin and function of symbols. Our symbols may function not just as embodiments of abstract linguistic ideas but also as conveyors of much deeper preverbal emotional truths that spring from the source of all emotions: families. As we will see in the next chapter, the structure of family relationships has exerted a profound influence on the subsequent shape and structure of human information systems. Relationships between parents and children, siblings, and extended family members form a psychological template that resurfaces over and over again in the way we seem to structure the relationships between thoughts and ideas, providing an implicit structure for the earliest taxonomies.

This emotional dimension of symbols may offer a partial explanation for why certain kinds of information systems prove more successful than others. Systems based on artificial ideologies, which lack a reinforcing emotional power, routinely fail to persist. For example, when communist regimes attempted to rewrite their countries' histories in a new ideological light, those efforts eventually faltered in the face of their citizens' own shared memories. "Because families continue to transmit to their children a more basic nonverbal presymbolic emotional 'truth,'" Greenspan argues, "these symbolic fabrications vanish immediately upon the collapse of the dictatorships that enforce them."[23] If our capacity for symbolic abstraction really rests so deeply in our limbic system, that may well explain why our most resilient information systems—like folktales, urban legends, and religious traditions—seem to flourish by passing through the strong social bonds of personal relationships rather than relying solely on institutional power or ideologies. And as we will see later in the book, oral traditions have proved particularly durable over the years, usually outlasting more elaborate written systems that come and go with the rise and fall of their institutional sponsors.

So far, we have seen how the deep biological history of networks and hierarchies has given rise to the escalating complexity of nature's information systems. We have seen how that process has equipped other species with the ability to preserve information beyond the life span of the individual organism through social imitation and pooling and by encoding memes onto their physical environments. And we have seen how, among primates, deep-seated emotional responses provided the semantic scaffolding for the emergence of symbolic expression—all of which brings us to the brink of human culture.

The next chapter explores how this long-standing interplay between hierarchical and networked systems ultimately yielded humanity's primal information system: taxonomy.

FAMILY TREES AND THE TREE OF LIFE

Civilization begins with a rose.

—Gertrude Stein

For many of us, the word *taxonomy* probably conjures images of sheer academic drudgery: obscure Latin names, plodding textbooks, and pallid biologists discussing the finer points of flora and fauna. Few nonbiologists are ever likely to give the subject much thought past high school, and even among many biologists, taxonomy has fallen into disrepute as a dismal backwater of the life sciences. But for the greater part of the past hundred thousand years, the practice of biological classification ranked as one of humanity's most essential cultural pursuits. "Some people dismiss taxonomies and their revisions as mere exercises in abstract ordering," writes Stephen Jay Gould, "a kind of glorified stamp collecting of no scientific merit and fit only for small minds that need to categorize their results. No view could be more false and more inappropriately arrogant. Taxonomies are reflections of human thought; they express our most fundamental concepts about the objects of our universe."[1] Taxonomies have played a crucial, and usually overlooked, role in shaping the structure of human thought.

What exactly is a taxonomy? A taxonomy, in its simplest form, is a system of categories that people use to organize their understanding of a particular body of knowledge. The oldest and most familiar taxonomies involve plants and animals. And although most of us may spend little time thinking about taxonomies, we use them every day to make sense of the world around us. For example, take a common animal like a brown long-haired tabby cat. Most of us would agree that a tabby is a kind of cat, that cats are mammals, and that mammals belong to the larger category of animals. So the taxonomy of a brown long-haired tabby cat might look something like this:

Animal
Mammal
Cat
Tabby cat
Brown long-haired tabby cat

Now, a biologist would consider this particular taxonomy woefully inadequate. The formal Linnaean taxonomy that most of us studied in some long-ago biology class uses a more detailed system that allows for up to nine levels of categorization, each accompanied by a signature Latin name. So, using the example above, the equivalent Linnaean taxonomy would look like this:

Kingdom: *Animalia*
Phylum: *Chordata*
Class: *Mammalia*
Order: *Carnivora*
Family: *Felidae*
Genus: *Felis*
Species: *Catus*

In this system, the long-haired tabby is a subspecies of the species *Felis catus*. Although the Linnaean system may boast of scientific accuracy, for most of us it does not mean much. The first group of categories, even though it might not pass scientific muster, nonetheless represents the kind of informal "folk" taxonomy that most of us rely on every day to make sense of the world around us, by slotting things into categories using everyday language.

For tens of thousands of years, human beings have been doing just this: giving names to the things around them and sorting those things into categories ("mammals," "birds," and "fish"). The oldest taxonomies stretch far back into our species' past—and perhaps even further into our pre–*Homo sapiens* lineage (although this possibility inevitably remains a matter of speculation). Indeed, our ability to classify things may be one of our species' great evolutionary differentiators. Human beings are the planet's uncontested master classifiers. Who else but Adam could God have chosen to name the animals?

While conventional histories of science often credit scientific pioneers like Carl Linnaeus and Aristotle as the first taxonomists, in truth taxonomies have a much longer history. The earliest folk taxonomies almost certainly predate literate civilization by many thousands of years. Even though we may never know precisely how our ancient forbears classified the natural world, we can surmise a great deal about how these systems may have worked from the study of modern-day tribal societies. Every human culture ever observed has created its own taxonomy of plants and animals.

Why are taxonomies so universal? Our capacity for classification seems to spring from two basic cognitive capabilities: binary discrimination (the ability to tell one thing from another) and lateralization of the human brain (the ability to string thoughts together). Once we distinguish two objects from each other, we can hold the objects in our minds long enough to recognize a conceptual distinction between them. This is how we distinguish black and white, male and female, good and evil, us and them. These basic categories in turn give rise to higher-level distinctions. For example, we can recognize the difference between a deer and an antelope, yet we can also recognize the ways in which they are alike. This ability to contrast degrees of sameness and difference provides the conceptual foundation for hierarchical thinking. So we can recognize that a deer and an antelope are simultaneously different (belonging to separate species) and alike (belonging to a larger family of species that also includes bison but does not include, say, rabbits). Working our way up through levels of similarity between animals, we might eventually come up with something like "mammals." Our prehistoric ancestors spent a great deal of time doing just this, creating categories to describe the living things around them. "Our predisposition to classify at all is an ancient trait," writes the anthropologist Brent Berlin, "and clearly has an adaptive advantage."[2]

Human cultures all over the world appear to have developed surprisingly similar strategies for categorizing plants and animals—from the tribes of New Guinea, Australia, and South America to the elaborate layered taxonomies of Confucianism, Islam, and the European Renaissance. Although the details of individual classification systems vary widely, the structure of these taxonomies reveals remarkable similarities. Something in our cognitive makeup seems to drive us to categorize the natural world in hierarchical terms. Although we can never know exactly what our ancient ancestors' taxonomic systems may have looked like, we can make informed speculations by contrasting the taxonomies of modern tribal cultures with traditional European systems that almost certainly contain traces of earlier prehistoric systems.

Berlin has performed extensive field studies on folk classification systems among modern tribal communities. Studying the Aguaruna tribe of Ecuador, Berlin has noted that the tribe members' appreciation of the tropical forest is nothing less than phenomenal. "Walking through the tropical forest with an Aguaruna guide is an awe-inspiring experience. One is quickly provided with a separate name for what appears to be each botanically distinctive tree.... The group recognizes at least fifty distinct classes of palms."[3] Folk taxonomies are often startlingly complex. "To find 'simple savages' controlling an extensive body of knowledge akin to the scientific fields of botany and zoology," writes Berlin, "is truly remarkable."[4]

Throughout our species' time on earth, we have relied for survival on collecting and disseminating accurate information about the world around us: knowing which animals we can eat, which plants are poisonous, and so forth. For humans living in hunter-gatherer or tribal societies, folk taxonomies are fundamental tools of group survival. It would stand to reason, then, that natural selection played some role in the formation of these systems. Individuals born with an aptitude for taxonomy would likely stand a higher chance of reproductive success. If you could remember which snakes were poisonous, which mushrooms you could eat, which animals you could tame, you stood a better chance of passing on your genes to the next generation. No single human being could possibly catalog the entirety of nature's creation from scratch; it would take too much time and a great deal of trial and error (often of the life-ending variety). Folk taxonomies took generations to build.

What can folk taxonomies teach us about preliterate human communities? The anthropologist Cecil Brown conducted the first comprehensive study of how preliterate human social groups organize their knowledge of the animal world. Surveying a broad range of folk classifications across numerous language groups, Brown sought to determine the level of agreement between geographically disparate cultures. The parallels, he discovered, run surprisingly deep. Even the most seemingly divergent cultures seem to employ almost identical strategies for organizing information, following a pattern that Berlin dubbed ethnobiological rank, or

TABLE 1

	NAME	DESCRIPTION	EXAMPLES		
1	Unique Beginner	The highest level of taxonomic inclusion	Plant	Animal	Computer
2	Life-Form	The first order of division, always "polytopic" (consisting of at least two members)	Tree	Mammal	Apple Computer
3	Generic	A "psychologically primary" category, usually identified with a unique name	Oak tree	Dog	Macintosh
4	Specific	A secondary name, usually involving a qualifier added to the generic name	White oak tree	Hound dog	Mac Mini
5	Varietal	The final level of granular description	Swamp white oak tree	Basset hound	Mac Mini G4 1.42 GHz
	Affiliate	A horizontal or "meta" category that might include members of multiple genera	Deciduous trees	Pets	Personal computers

the use of hierarchical categories to describe the characteristics of plants and ani-mals. According to Berlin, every tribal community ever studied appears to share a universal tendency to divide their knowledge of plants and animals into five or six nested categories:[5]

Not only is a rose a rose, then, but a flower is a flower, a plant is a plant, and so on. All human cultures appear to share the habit of categorizing flora and fauna using a similar structure; they also reach surprising degrees of consensus on the substance of those categories. Among the most commonly observed biological groupings are trees, small plants, bushes, vines, and grass. In the animal world, the most common zoological classifications include bird, fish, snake, "wugs" (what we colloquially call creepy-crawlies: small creatures that always include insects, often spiders, and sometimes worms), and mammals. And the similarities go even deeper. Every culture also seems to agree that things have "real names"—that a rose is more a "rose" than it is a "plant." This psychologically primary category al-ways lands squarely in the middle of the taxonomy, suggesting that not only do all cultures share a disposition toward creating nested taxonomies but that they also seem to recognize one category as being more accurate or true than the others.

Why do people around the world seem to reach such similar conclusions about the natural world? Surely, there is nothing inherent in the plants or animals them-selves. As Charles Darwin later proved, Aristotle's notion of idealized forms was classical wishful thinking. We can only assume that our innate disposition toward the animal kingdom reflects a kind of evolutionary consensus. How can we ex-plain the remarkable consistency of these information systems across far-flung cultures? There is no gene, as far as we know, that dictates the structure of taxono-mies, yet all human beings seem to share a certain disposition toward categorizing and naming things in structurally similar ways. Just as we perceive similar degrees of the color spectrum or a need to tell stories to each other, we seem to have a deep inborn urge to categorize the world around us.

The evolutionary advantage of taxonomic consensus seems self-evident. Per-haps we can imagine our distant ancestors huddled around a fire planning the next day's hunt; surely, it would benefit the group as a whole if everyone could agree on a goal of hunting "bison" as opposed to, say, "mammals." Similarly, re-productive advantage would accrue to those people who knew to tread carefully around "snakes," rather than worrying about "reptiles" in general, or making fine-tooth distinctions between green snakes and rattlers. The psychological primacy of the genus seems to make eminent evolutionary sense.

The remarkable similarities between folk taxonomies suggest that categori-cal thinking cannot be the product of human cultural idiosyncrasy. The paral-lels simply run too deep; there must be an evolutionary basis. Just as all human beings are born with a disposition toward spoken language, so we also seem to share an affinity for taxonomic thinking. Although the contents of those categories differ from culture to culture, all human beings seem to be born with a tendency to understand the world in terms of nested categories.

The notion that one of humanity's most essential information systems may have a genetic basis might come as troubling news to cultural relativists as well as technological idealists who insist on proclaiming the imminent death of hierarchical systems.

Prototype Theory

If evolution has equipped us with a disposition toward recognizing things as having "real names" at a certain level in a taxonomic hierarchy, then it stands to reason that such a disposition might continue to reverberate even after human beings moved on from thinking about plants and animals to consider a larger body of subject matter. The structure of folk taxonomies seems to have shaped the way human beings create all kinds of categorical meaning. In his classic 1958 essay "How Shall a Thing Be Called?" the philosopher Roger Brown observed that "the dime in my pocket is not only a dime. It is also money, a metal object, a thing, and, moving to subordinates, it is a 1952 dime, in fact, a particular 1952 dime."[6] Yet somehow, most of us would likely call a dime a dime.

Brown's intuition meshes exactly with Berlin's model of ethnobiological rank, suggesting that our ancient taxonomic tendencies influence our perceptions of other kinds of categories as well. According to Brown, the dime is the thing's "real name," reflecting the same notion of psychological primacy that Berlin identifies with the generic level of categorization. Linnaeus too seems to have recognized the primacy of the genus, placing it at the heart of his own classification system. Berlin's research into folk classification reveals the same result: that the genus occupies a central position as the "real" name of any given organism. The genus represents a kind of focal point in our orientation to the world from which all the other categories derive.

In this universal human tendency to organize information into categories, we can also recognize the familiar contours of networks and hierarchies at work. A folk taxonomy is a classic hierarchical system, but it depends for stability on the reinforcing presence of a network. Ontological hierarchy provides a system capable of scaling to encompass the phenomenal world, while the underlying social network cements its role in the culture and ensures its longevity from generation to generation. This archetypal pattern of hierarchical taxonomy reinforced by social networks echoes repeatedly throughout the history of humanity's subsequent information systems. Recent research into human-computer interaction seems to back up the claim as well. In 1996, IBM researchers discovered that when web users tried to navigate their way around the IBM website, they become lost once they traveled beyond five or six levels deep in the menu structure, even though the site comprised a series of menus extending fourteen or fifteen levels down. Perhaps at some unconscious level, our experience with the web owes something to our evolutionary disposition toward understanding the world in terms of taxonomic category structures.

Although we may never be able to model the evolutionary basis for taxonomic systems in strictly biological terms, the presence of such rules would suggest that the structure of our modern classification systems—with their hierarchical categories, binomial naming schemes, and cross-referencing—are not the product of individual human genius (which is not to denigrate the contributions of great taxonomists like Aristotle and Linnaeus) but that these systems emerged over a much longer period stretching deep into our genetic past. If taxonomy is rooted in our evolution as a species, the structure of these systems should also have a corresponding cognitive basis in the human mind. And indeed, the human mind appears to be particularly well suited to think in terms of hierarchical categories. Here, we must refer briefly to the work of prototype theorists, who have pioneered a new understanding of how the mind creates categories.

The psychologist Eleanor Rosch pioneered prototype theory, a breakthrough model of human cognition that predicts how the mind assigns things to categories not only based on cultural information but through a process of direct sensory engagement with the phenomenal world. Her theory suggests that people form categories based on "prototypes," or best examples of a particular kind of thing. The theory explains, for example, why people in different cultures tend to recognize similar types of colors or similar kinds of animal species. For ex-

FIGURE 3. © Alex Wright, 2022.

ample, consider the following pictures of birds. If you had to select one of the images below as the best example of a "bird," which one would you choose?

Chances are, you would pick the second image as the best example of a "bird." The crane on the left is taller than most birds; the penguin on the right cannot fly; but the bird in the middle is similar in size and shape to most other kinds of birds, so it falls closest to the best example or prototype.

Prototype theory suggests that we have not only an innate disposition toward categories but also that we have an ability to identify levels of gradation within those categories. This capacity for gradation—that is, the ability to recognize that things may be both different and similar at the same time—holds the key to our capacity for hierarchical categorization. One-dimensional categories are inherently dissatisfying to the human brain; we seem to have an innate need to separate things into finer-tuned distinctions. As a result, we form nested categories to describe the relative similarities and differences between things. In all likelihood, folk taxonomies evolved in concert with our cognitive facility for hierarchical categorization.

Rosch's work has, alas, been widely misinterpreted. A simplistic reading of her research would suggest that categories are determined through perception alone—that is, that there are categories inherent in the world. Rosch herself even subscribed to this view in her early work, giving rise to much of the subsequent misinterpretation. But she later revised her theory, coming to the conclusion that prototypes themselves do not determine the semantic categories that people formulate; rather, these are determined through a more complex process of cultural transmission. "Prototypes do not constitute a theory of representation for categories," she wrote. "The facts about prototypes can only constrain, but do not determine, models of representation."[7] In other words, the categories themselves are transmitted through culture, but the human brain is born ready to receive such transmissions.

In his landmark book *Women, Fire, and Dangerous Things*, the linguist George Lakoff wrote a sweeping refutation of the classical (or objectivist) view of categories. The classical view stems from the Aristotelian notion of things having intrinsic characteristics, ideal forms. It is predicated on the possibility of universal truth based on the observation of derived properties. This approach served as the foundation of the modern scientific method and Linnaean taxonomy; it would later run afoul of findings in evolution. Classical categories proceed from disembodied ideals and support the notion that ideas inhabit another plane of thought.

> To question the classical view . . . is thus to question the view of reason as disembodied symbol-manipulation and correspondingly to question

the most popular version of the mind-as-computer metaphor. Contemporary prototype theory does just that—through detailed empirical research in anthropology, linguistics, and psychology. . . . Human categorization is essentially a matter of both human experience and imagination—of perception, motor activity, and culture on the one hand, and of metaphor, metonymy, and mental imagery on the other. As a consequence, human reason crucially depends on the same factors. And therefore cannot be characterized merely in terms of the manipulation of abstract symbols.[8]

Our capacity for categorization, in other words, has a biological basis in terms of both individual cognition and our larger social interactions. Prototype theory also meshes seamlessly with E. O. Wilson's theory of gene-culture coevolution. The human brain appears to have evolved in concert with the social transmission of categorical information. As a result, we are all born with a deep-seated need to understand the world in terms of categories and to share that understanding with each other.

Beyond Flora and Fauna

Our long heritage of taxonomic thinking has influenced more than just the way we label plants and animals. For tens of thousands of years, these archetypal information systems weaved themselves into the fabric of human social life. Folk taxonomies became deeply entwined with the first mythologies, taking on progressive layers of meaning that became embedded in our cultural heritage. In 1903, the sociologist Émile Durkheim, along with his colleague Marcel Mauss, published a pioneering study of the larger social role of folk taxonomies titled *Primitive Classification*. In one of the first studies of its kind, Durkheim surveyed the taxonomic practices of Aboriginal peoples in Australia and North America. His research led him to probe the little-understood relationship between classification systems and human social organization.

Durkheim posited that the structure of classification systems hewed closely to the structure of existing family and kinship networks, expressing a deep unconscious need for members of tribal societies to project their otherwise inscrutable kinship relationships onto the outside world. When people described the natural world, therefore, they did so by invoking the language of family relationships: plants and animals belonged to "families." The fir tree was the child of the category "tree," the sibling of the pine, and perhaps a distant cousin of the bush, but it was no relation to a chicken. Tribal conceptions of the natural

FIGURE 4. Bach Family Tree, ca. 1750–70. Photo credit: Bridgeman Images.

world paint the animal kingdom as one big interlocking web of families. Thus, natural classification systems reflected existing kin relationships; the vastness of the natural world provided a kind of conceptual mirror that allowed tribe members to ponder the larger web of connections among themselves.

In Australia, Durkheim observed how Aboriginal tribes classified the natural world using a system that meshed seamlessly with their own tribal organization. Each tribe was divided into "moieties," or social units, each of which in turn was subdivided into so-called marriage classes and further subdivided into smaller clans. Each of these clans, in turn, claimed an affiliation with some natural element: the trees, the plains, the sky, the stars, the wind, the rain. Each clan also followed certain dietary restrictions according to their marriage class (members of one class ate opossum, kangaroo, dog, and honey, while members of another class ate emu, bandicoot, black duck, and certain snakes). Like the Zuni people (see chapter 1), the Aboriginal tribes used their own family structures as living containers for categorizing information about plants, animals, and the natural elements.[9]

Looking deeper into the symbolic relationships between family organizations and the associated animals and elements, Durkheim discovered the correlation between family subdivisions and the ethnobiological ranking systems. "In fact the moiety is the genus," he wrote. "The marriage class is the species; the name of the genus applies to the species," and so forth. "We are no longer dealing with a simple dichotomy of things," wrote Durkheim, "but with hierarchized concepts." For these remote tribal societies, Durkheim came to the conclusion that the structure of early human classifications constituted an externalized representation of macro social structures. The observed hierarchies of the natural world, in Durkheim's view, represent a projection of humanity's own social relationships. The two systems reinforce each other—indeed, in the "primitive" mind, they are indistinguishable.

The first categories may have emerged as outward expressions of existing social relationships: an attempt to make visible otherwise invisible relationships like "parent," "sibling," or higher-level abstractions like "family." As Michael Hobart and Zachary Schiffman write, "Genealogy provides the ideal classificatory tool, for it narrates a sequence of actions. It thus sustains the tradition while, at the same time, subjecting it to a hierarchical ordering that clarifies the nature of various figures. When gods are considered, genealogy becomes a means of understanding the cosmos. . . . When mortals are considered, it becomes an encyclopedic framework for historical and geographical as well as social information."[10]

Durkheim believed that the tendency toward categorical hierarchy simply reflects the natural hierarchy of family structures, thus explaining the prevalent human tendency to describe groups of plants and animals in kinship terms. The family tree, in other words, works as a two-way metaphor: families are shaped like trees, and trees are shaped like families. "The first logical categories were social categories; the first classes of things were classes of men, into which these things were integrated. It was because men were grouped, and thought of themselves in the form of groups, that in their ideas they grouped other things, and in the beginning the two modes of grouping were merged to the point of being indistinct." So in Durkheim's view, categories derived from existing social structures, forming a conceptual template into which the tribes poured their knowledge about the natural world. "Moieties were the first genera; clans, the first species. . . . It is because human groups fit one into another—the sub-clan into the clan, the clan into the moiety, the moiety into the tribe—that groups of things are ordered in the same way."[11]

If tribal classifications do indeed represent externalized projections of social relationships, it seems fair to surmise that these systems must also entail some level of emotional investment. These systems succeeded because they tapped into the reinforcing power of human emotion by mirroring existing family relationships. If we accept Stanley Greenspan's assertion that higher-order information exchange relies on emotional triggers in our limbic system (see chapter 1), then we could accept Durkheim and Mauss's assertion that "a species of things is not a simple object of knowledge but corresponds above all to a certain sentimental attitude."[12] Berlin, however, questions Durkheim's conclusions about the social origins of taxonomies: "The ethnobiological data suggest strongly that [Durkheim and Mauss] were wrong in their proposal on the social origins of classification. My strong intuition is that it was just the other way around; knowledge of natural kinds and the similarities and differences between and among species provided the model for the social classifications that humans ultimately constructed to provide meaning to social relations."[13] In other words, perhaps people did not create taxonomies to mimic the structure of social relationships; instead, they modeled their social relationships after their observations about the natural world. Regardless of which came first—the human family or the taxonomic one—there is no question that taxonomies and human social structures evolved in concert with each other.

Multilayered, socially pregnant taxonomies would eventually find a complex and beautiful expression in advanced mythological systems like those found in Greece, China, and elsewhere, where systems of astronomy, astrology, geomancy, and divination merged into complex, highly stylized classification systems with deep taxonomic roots. The Chinese folk classification centers on the division of space into four cardinal points, each of which is further divided in two, resulting in eight compass points that, like the Zuni system (see chapter 1), represent the natural elements: heaven, earth, clouds, fire, thunder, wind, water, and mountains. For example, the south is associated with the father principle as well as with sky, pure light, force, the head, heaven, jade, metal, ice, red, and horses. In opposition to the fatherly south stands the motherly north, associated with earth, darkness, docility, cattle, the belly, and cloth, among other things. Thus, the basic genealogical principle of mother and father expands into a broad classification system framing the Chinese people's understanding of the world around them. Each direction is also associated with a season (each of which is further subdivided into six, making for the twenty-four seasons of the Chinese calendar), an animal (the dragon to the east, red bird to the south, white tiger to the west, and black tortoise to the north), a color, an element, and certain underlying principles.[14]

The classification of natural phenomena into eight categories resembles the strategies employed by other human cultures that developed oceans away: the Australian Aborigines, Zunis, and numerous other human societies have developed analogous systems relying on the correlation of cardinal directions with seasons, animals, elements, and genealogical relationships. Humanity's inborn disposition toward hierarchical systems seems to manifest over and over again in these ancient information systems, taking on increasingly complex expressions as human social groups expand in size. Could the striking similarity between these systems suggest the further operation of evolutionary rules at work?

The Chinese astrological calendar also has deep taxonomic roots, connecting each year in a twelve-year cycle with a totemic animal: rat, cow, tiger, rabbit, dragon, snake, horse, goat, monkey, rooster, dog, and pig. The animals divide into four groups of three, each associated with a compass direction and with one of four natural elements (fire, water, earth, wind). Thus the entire classification system pulls together an integrated worldview, encompassing time, geography, the animal kingdom, and elemental forces. As Durkheim and Mauss put it, "Here we have, then, a highly typical case in which collective thought has worked in a reflective and learned way on themes which are clearly primitive."[15] The guiding principles of Chinese classification provided the philosophical framework guiding everything from architecture (as in the well-known practice of feng shui) to affairs of state, personal relationships, and the Chinese conception of the natural world. Over a period of several thousand years, Chinese civilization became a living manifestation of a vast and all-encompassing classification system. The Chinese system provides perhaps the most compelling example of an ancient classification that survived and adapted itself through long periods of upheaval: wars, empires, and the emergence of a vast institutional bureaucracy.

On the other side of the world, the people of ancient Greece were developing an astrological and divinatory system conceptually not far removed from the Chinese system, predicated on relationships between space, time, and the elements. Here again, the system began with the correlation of natural elements with certain directions, colors, and natural phenomena, all framed in terms of genealogical relationships—in this case, between the gods. For example, Poseidon, the Greek god of the seas, was the brother of Demeter, the goddess of agriculture. Artemis, the moon goddess, was the sister of Apollo, the god of light. Ares, the brutal god of war, was brother to Athena, the goddess of wisdom, justice, and the arts. Each god, in turn, represented a set of naturally occurring phenomena. Often, the gods would appear under alternate guises. "Every god has his doubles," writes Durkheim, "who are other forms of himself, though they have other functions; hence, different powers, and the things over which these powers are exercised, are attached to a central or predominant notion, as is the species to the genus or a secondary variety to the principal species."[16] Like other mythological systems, the Greek pantheon employed the familiar construct of family relationships to invoke a web of semantic relationships (see chapter 5 for a further discussion of Greek mythology).

In human cultures throughout the world, ancient preliterate mythologies reveal this pattern of gods and other supernatural beings as coded entities, organized into families, who represent the phenomenal world. The oldest mythologies, with their divine genealogies, echo the familial structure of folk taxonomies, forming a hierarchical scaffolding to allow preliterate peoples to organize their knowledge about the natural world. As the pantheon of gods and mythic relationships grew over time, specialized gods began to cluster into increasingly narrow categories, until finally the smaller gods became mere echoes or alternate names of their former selves. With the advent of writing, many of these mythologies would give way to a more empirical understanding of the world. As these early classification systems grew in complexity, along with the human social structures that created them, they also laid the groundwork for the gradual erosion of superstitious belief. As Durkheim and Mauss put it, "Mythological classifications, when they are complete and systematic, when they embrace the universe, announce the end of mythologies."[17] Even in our highly systematized age, however, we still feel the reverberation of old mythologies, the ancient archetypes of the classifying mind.

THE ICE AGE INFORMATION EXPLOSION

We are symbols, and inhabit symbols.

—Ralph Waldo Emerson

In ancient Peru, Incan messengers used to travel across the Andes carrying a bundle of woven thread known as a *quipu*, or "talking strings." When a messenger arrived at his destination, he would deliver his news while reeling off knots in the string like a rosary. For the Incas, a people with no written language, the quipu served as their core information technology: it was a newspaper, a calculator, even a repository of laws. A skilled *quipucumaya* ("keeper of the quipus") could use the device to tell complex stories by weaving the colored threads together. Each thread represented a different facet of the narrative: a black string marked the passage of time, while other colored strings symbolized different characters or themes in a story: rulers, neighboring tribes, gods.[1] By juxtaposing the multicolored strings of the story along the black-stringed axis of time, the quipucumaya could "write" almost any kind of narrative. Despite a total lack of writing as we would understand it, the Incas managed to keep track of enormous stores of information by manipulating these symbolic objects.

Until recent years, historians tended to give nonliterate cultures short shrift as systems thinkers. In the popular caveman stereotype, prehistoric people were cultural simpletons, too preoccupied with survival to pursue the loftier avenues of civilization. Although mainstream academic views have shifted to accommodate a growing appreciation for preliterate societies, our mainstream culture still harbors a vague prejudice against people who do not know how to write. As a result, we often fail to appreciate the richness and complexity of nonliterate cultures and what they can teach us about the way our minds work. As it turns out, preliterate societies not only exhibit surprisingly complex cultural behav-

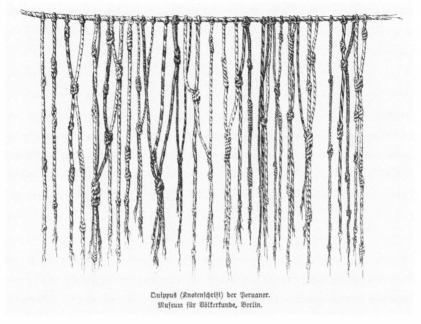

Quippus (Knotenschrift) der Peruaner.
Museum für Völkerkunde, Berlin.

FIGURE 5. Quipu. From *Illustrierte Geschichte der Weltliteratur*, by Dr. Johannes Scherr (Franckh'sche Verlagsbuchhandlung, Stuttgart, ca. 1895).

iors but in many cases make innovative use of symbolic information technologies to record, store, and distribute their shared knowledge.

Although symbolism may seem like an intrinsically human capacity, the practice of using external symbolic artifacts emerged only in our relatively recent past. *Homo sapiens* reached anatomical modernity at least one hundred thousand years ago, yet it took us more than sixty thousand years to begin producing the kinds of symbols that most of us recognize as the manifestations of human culture. Although our latent capacity for symbolism may have taken shape deep in our primate lineage (see chapter 2), that ability seems to have lain dormant until comparatively recently in our species' history. For most of human beings' time on earth, we have behaved in ways barely distinguishable from our Neanderthal cousins: foraging in loose-knit groups; fashioning rudimentary tools like sticks, rocks, and spears, occasionally burying our dead. Then, about thirty thousand to forty thousand years ago, our ancestors suddenly started displaying a whole tapestry of symbolic behaviors: decorating bone knives, stringing bead necklaces, making paintings. "In a blink of geological time," as the science writer John McCrone puts it, "one line of hominids suddenly became symbolically minded and self-aware."[2] That we should have survived for so long without producing symbolic artifacts is hardly surprising. After all, every other species on earth has managed to survive without producing a single bead necklace, cave painting, or pop music CD. The question that still vexes many researchers is this: Why did we cross the threshold to symbolic expression so quickly? Human mythologies are rife with explanations for the emergence of symbolic thought: Eve biting the apple, the appearance of a primordial Buddha, or the arrival of mystical space aliens, to name a few.

There is also a more mundane explanation, one grounded squarely in the archaeological record. The arrival of symbolic expression appears to have coincided almost exactly with a period of rapid climate change, starting about forty-five thousand years ago, when the last great Ice Age sent glaciers out across much of the earth's surface. As temperatures plunged, our ancestors started competing over dwindling food supplies. Competition for resources brought people into closer social proximity, and that proximity appears to have triggered the emergence of symbolic expression. Call it a primordial tipping point.

In the popular stereotype of "primitive" or preliterate societies, we might think of our ancestors as big game hunters from time immemorial, but in fact we come from a long line of scavengers. For our first sixty-thousand-odd years, *Homo sapiens* were scrappy opportunists who foraged for plants, chased foxes and beavers, and lunged after rabbits and pigeons. As global cooling sent temperatures plummeting, however, those once-easy pickings ran scarce, and our ancestors had no choice but to set their sights on, literally, bigger game.

Now, any able-bodied biped can thwack a pigeon, but stalking a woolly mammoth or an eight-foot-tall prehistoric cow takes teamwork. Faced with the choice between starvation and banding together to hunt larger prey, humans started working together. Whereas once they had moved in loose-knit bands with little or no social structure, now they started collaborating in larger groups, hunting big animals, huddling in caves for warmth, and forging the kinds of social bonds necessary to support an expanding body politic. Social relationships grew more complex, demanding new forms of mediation. And somewhere along the way, people started making symbols.

For Ice Age humans, symbols appear to have emerged only after we came in from the cold and started working together. The first unambiguously symbolic objects were beads and pendants, usually made of stone, shells, or ivory. These objects would have been worthless as tools; their only plausible function had to involve some form of symbolic expression. Although we can never know exactly what these objects meant to people, archaeologists comfortably speculate that beads may have served as emblems of social identity, marking affiliation with a particular band or tribe, denoting gender, or marking major rites of passage like coming of age or marriage. The use of these totems seems to suggest the value in having a means to identify oneself and one's status in a larger social group. In smalls group, people can easily recognize each other as individuals. The larger a group becomes, the harder it gets for everyone to know each other. In lieu of face-to-face recognition, people started relying on external signifiers as a way to communicate meta-information about themselves: social rank, marriage status, and so forth. As Ice Age humans began to shift their way of life from loose-knit bands to larger tribal communities, they began to build what Steven Kuhn et al. call ornament technology, using a lattice of symbols with shared meaning to signal status and form the basis for new kinds of relationships.[3] The use of beads and other totems facilitated what Clive Gamble calls a "release from proximity,"[4] a rudimentary kind of social contract built around a shared understanding of what certain symbols meant. With an agreed set of symbols in place, human groups could begin forging wider networks of trust. And from those networks, new kinds of hierarchies emerged.

During the Upper Paleolithic period, starting about thirty-five thousand years ago, people seem to have devoted enormous time and energy to making beads of astonishing sophistication, with painstaking attention to detail. In one European Ice Age community, bead making seems to have turned into a mass obsession, with over fourteen thousand beads appearing at one twenty-eight-thousand-year-old burial site. "These beads were very important to these people," says the New York University archaeologist Randall White. "They spent thousands of hours of their life making ivory beads even though they were living within 150 kilometers of a huge glacial ice sheet."[5]

Beads seem to have functioned not only as communication objects within particular groups but also as units of cultural exchange between otherwise disparate groups. At the caves of Altamira in Spain, archaeologists have discovered an enormous variety of symbolic objects—ostrich-shell beads, ivory figurines, seashells, pieces of amber—suggesting that the site may have served as a trading post for Ice Age Europeans: a hub on the regional trading network and a nexus for a kind of protocommercial exchange. The sheer variety of styles suggests that different social groups gathered there to exchange these artifacts (and probably other goods as well) as a means of solidifying social relationships or forming alliances. These early trade networks served much the same purpose as the Internet today: as a vehicle not just for commercial exchange but also for forging social bonds between far-flung communities.

The emerging technology of symbolic communication would wreak lasting changes on the structure of social organization. Symbolic communication provided a networked communications platform that allowed more complex social, and eventually political, hierarchies to emerge. The relationship between symbolism and political power stretches back at least thirty thousand years. Given that human beings survived so long without symbols, however, why has symbolic

FIGURE 6. Necklace, beads, and pendants, El-Wad Cave, Mount Carmel, Natufian culture, ca. 11000 BC (shell and bone). Collection of the Israel Antiquities Authority / Photo © The Israel Museum, Jerusalem by Meidad Suchowolski / Bridgeman Images.

culture proved such a durable adaptation? As we have seen in the previous chapters, symbols are more than just semantic emblems; they rely on emotional responses lodged deep in our limbic systems. Understanding this emotional aspect of symbolic expression sheds light on why some systems of symbolic expression persist while others fail. Effective symbols trigger deep-seated emotional responses. By triggering limbic system responses, simple totems like beads and pendants allowed people to transfer emotional bonds from people to objects; in turn, the objects could function as charged particles of emotional meaning, enabling people to forge increasingly far-flung ties with each other. In Ice Age communities, beads and pendants were imbued with emotional meaning that could invoke responses between strangers grounded in the deeper emotional bonds of kinship: power, loyalty, or even something like love.

Today, we still rely on certain kinds of symbols in much the same way as our Ice Age forebears. The use of symbols as tokens of social trust is evident across the World Wide Web, where people routinely rely on disembodied symbols to evoke emotional responses and forge bonds of trust with each other. At a simplistic level, people signal emotions to each other through the widespread use of icons in instant messaging applications to represent emotional states (smiley faces, angry faces, sad faces, etc.). At a more subtle level, web users rely on totemic symbols to facilitate further degrees of social abstraction. Consider the prevalence of trust ratings on eBay, where complete strangers routinely transact business with each other purely on the basis of iconic representations of trust. An eBay user's trust rating functions as a valuable form of social capital. As successive users exchange positive or negative ratings of each other over time, that information accumulates in the form of a trust rating. Some users begin to accumulate new levels of social rank embodied in totemic icons (e.g., "Power Seller"). Similar trust mechanisms play a crucial role in facilitating the success of user reviews at Amazon and any other number of websites where people routinely exchange information and seal bonds of trust with complete strangers. These interactions are not so far removed from the symbolic exchanges of trust and coding of social rank that Ice Age people achieved using beads, shells, and necklaces.

Symbolic expression, then, originated not just as an aesthetic or a spiritual pursuit but also as a pragmatic adaptation to changing environmental conditions. "Art is a cultural equivalent of mutation," says White. "We know that bipedalism was caused by a genetic mutation that allowed some creatures to be able to stand upright, and that proved advantageous. What about the individual or social group that creates drawings of animals? What is the adaptive value of that? Well, for one thing, it creates the ability to communicate with each other through images." From this perspective, we can view Ice Age art as more than just a first foray into spiritual or aesthetic expression but as a practical adaptation to changing environmental conditions. "We've tended to think of the imagery from this period as 'art' in the soft sense that people were into evocative images, and animals really meant a lot to them, etc.—and maybe they did. But you also have people who are capable, using artwork, of showing someone else how an animal behaves, or where best to plant a spear, or any number of practical applications."[6]

We often think of "art" as a rarefied form of human achievement, a capacity that emerges only during periods of relative prosperity. Abraham Maslow's famous hierarchy of needs places cultural expression at the very top of the pyramid, expressing the common assumption that humans come to worry about aesthetic expression only after we secure our supposedly "basic" needs like food, shelter, and physical security. But the archaeological record suggests that symbolic expression was no esoteric exercise; it was a matter of survival. Symbolic expression seems to have played a critical role in facilitating tighter working partnerships and in establishing functioning social structures to ensure group survival. Creating symbols with shared meaning turns out to be a useful way of cementing social relationships. As Émile Durkheim writes, "In all its aspects and at every moment of history, social life is only possible thanks to a vast symbolism." Indeed, the advent of symbols appears tightly woven to the emergence of complex social relationships.

White has studied the evolution of art among Ice Age cultures and has concluded that art played an essential role in enabling *Homo sapiens* to survive during this period. "People don't think about art as a necessity," says White. "But it seems to have arisen as an attempt to cope with some very difficult circumstances. Art corresponds to periods of stress much more than it corresponds to periods of well-being and leisure. There is no question that art is not being produced as a result of people having more spare time; it is being produced when all hell is breaking loose and people don't have enough to eat."[7]

The most spectacular examples of Ice Age art are the famous cave paintings of Cro-Magnon Europe. At the caves of Lascaux, Altamira, and a host of smaller sites, the charcoal-and-pigment paintings and etched figures of bison, mammoth, and even rhinoceroses have captivated the public imagination for years. Scholars have variously speculated that cave paintings were used for hunting rituals or religious ceremonies; more recently, some scholars have suggested they may be the ancient equivalent of teenage graffiti.[8] Although the purpose of these paintings inevitably remains subject to speculation, most scholars agree that the paintings served at least in part as a practical method for building social consensus. In all likelihood, cave paintings served more than just a ritualistic or an aesthetic purpose; they may also have functioned as informational tools, the preliterate equivalent of a how-to manual or encyclopedia. Cave paintings allowed social groups to coalesce around fixed images, shared reference points that gave them a means of aligning their thoughts toward a common object embodied in an exteriorized expression. Just as stigmergy provides a mechanism for indirect collaboration between individual members of a social group, so cave paintings may have served as a means for people to introduce social constraints on each other's behavior.

Although historians typically assign writing a special status as the harbinger of civilization, the building blocks of complex information systems were in place long before the first scribe set stylus to clay. "Each society creates culture and is created by it," writes E. O. Wilson. "Through constant grooming, decorating, exchange of gifts, sharing of food and fermented beverages, music, and storytelling, the symbolic communal life of the mind takes form, unifying the group into a dreamworld that masters the external reality into which the group has been thrust."[9] Human beings have invested long millennia in forging social bonds through the stigmergy of symbols, invoking our shared yearning to make our dreamworlds real.

The Ice Age information explosion brought humanity to the brink of literacy. For tens of thousands of years, the infrastructure of tribal information systems—folk taxonomies, mythological systems, and preliterate symbolism— created the conditions for the emergence of literate culture. For most of our species' time on earth, we relied on oral traditions—storytelling, folk taxonomies, and gesture—to preserve and transmit information between generations. The advent of externalized symbols would change the shape of human social organization, bringing people together in larger social groups and, eventually, lead them across the threshold to an even more revolutionary adaptation: writing.

THE AGE OF ALPHABETS

Littera scripta manet.

—Horace

Stories explaining the origins of writing vary as widely as the cultures that produce them. The Egyptians believed that Thoth, god of wisdom, created the first hieroglyphs. The Norse creator god Odin was said to have invented runes. The ancient Greeks credited Hermes with inventing the Greek alphabet. The Mayans attributed the art of writing to their creator god Itzamná. Certain Arab traditions maintain the first writer to have been no less a personage than Muḥammad. Indeed, almost every human society that has crossed the threshold to literacy has at some point attributed the advent of writing to some form of divine intervention.

Although traditional Western histories have typically situated Mesopotamia as the birthplace of writing, many historians now believe that writing emerged independently in at least three locations: Sumer, China, and the ancient Mayan civilization. In each case, however, the advent of writing seems to have come about in tandem with the growth of population centers and a corresponding increase in social, economic, and political intercourse. By about 5000 BC, a growing population in the ancient settlement of Uruk (in what eventually became the kingdom of Sumer) was coalescing into the first human settlement to resemble a city: "a mind-defeating jumble of temples, dwellings, storerooms, and alleyways," as Thomas Cahill describes it.[1] The people of Uruk, having mastered a few basic agricultural techniques, had started to settle in a growing population center with a flourishing trade and increasingly specialized classes of workers: farmers, merchants, builders, and a ruling class of elite citizens. It was the world's first boom town.

The sheer concentration of people created a cauldron of technological innovations. During this period, ancient Sumerians invented the wheel, sailing ships,

FIGURE 7. Thoth, inventor of writing, center, holding a pen and scribal palette. From the Book of the Dead of Hunefer (H. Judgement vignette with captions and spell 30. © Trustees of the British Museum).

molded bricks, metals, and architectural breakthroughs like vaults and domes. They were the first people to divide the day into hours.[2] Amid such a prolific period of innovation, the typical Sumerian might have scarcely noticed the curious little etchings that merchants were starting to make in clay tablets while they tried to keep track of their accounts. The booming commercial trade in Uruk created an unprecedented volume of transactions. As merchants struggled to keep up with an expanding commercial economy, they needed to develop better systems for keeping track of their accounts. The first forms of writing emerged "not for magical and liturgical purposes," writes V. Gordon Childe, "but for practical business and administration."[3]

Samuel Johnson once remarked that "no man but a fool ever wrote except for money." Indeed, if it were not for money, we might have no writing at all. Much as we like to think of writing as the fruit of high culture, all literature originates from the basest of all commercial transactions: sales receipts.

Writing appears to have emerged with little fanfare, arising as a practical innovation to take advantage of two prior technologies: drawing and counting. The earliest tabulating devices were notched bones dated to 35000–20000 BC— the earliest known representations of numbers. By the ninth century BC, clay tokens had emerged as units of basic currency—appearing almost simultaneously in Mesopotamia, Iran, Sudan, Palestine, and Syria. The first written notations were called *bullae* (tokens). Shaped like disks, cones, spheres, tetrahedrons, ovoids, cylinders, triangles, or animal heads, each token represented some form of commercial transaction: records of purchases, debts, or contractual obligations. The oldest specimens yet discovered date back as far as 3500 BC. By 3000

BC, the earliest examples of what we would recognize as "writing" emerged, when merchants began transcribing the contents of bullae onto clay tablets.

As written tablets took hold as a commercial technology, a new class of professional scribes emerged. These early knowledge workers facilitated trade by serving as trusted third parties who could record transactions between two or more parties. A typical written receipt might read something like this: "Offer Gentleman B the enclosed message from Gentleman A."[4] After kneading wet clay

FIGURE 8. Administrative clay tablet in cuneiform. Credit: Louvre, Paris, France © NPL—DeA Picture Library/Bridgeman Images.

into a cylinder or tablet, the scribe would record the details of the transaction by scraping marks into the surface using a wedge-shaped stylus. Both parties would put their marks on the tablet. The scribe would place his own seal, and the contract would be baked into the permanent record. Because clay dried quickly, the scribes had to work fast; this was no medium for poetry. Once the tablet was baked, the scribe would then deposit it on a shelf or put it in a basket, with labels affixed to the outside to facilitate future search and retrieval.

In China, the first forms of writing evolved from the practice of divination using so-called oracle, typically using the remains of oxen or turtles inscribed with sacred marks that would allow diviners to answer questions about the future. Although the earliest clear-cut examples of Chinese writing date to around 1200 BC, archaeological evidence suggests that prototypical versions of

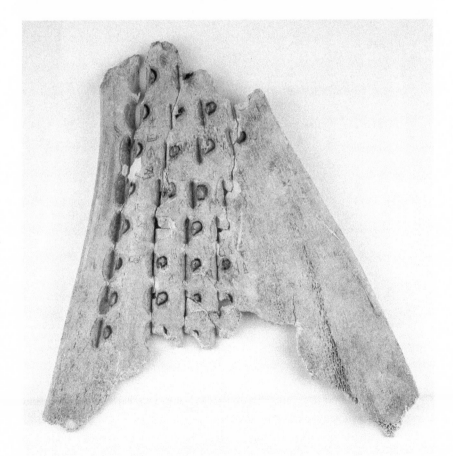

FIGURE 9. Chinese oracle bone, ca. 16th–10th century BC. © British Library Board (Or. 7694).

FIGURE 10.   Maya carved glyphs. Photograph by Gary Lee Todd. Maya Gallery, INAH, National Museum of Anthropology, Mexico City.

Chinese written characters may have started to appear as early as 6600 BC (but this remains a subject of scholarly debate).[5]

The Mayan hieroglyphic writing system appears to have emerged around 300 BC, but its roots may stretch further into prehistory into earlier pictographic systems that emerged in ancient Mesoamerican civilizations like the Olmec. By the third century AD, the Mayans had developed a full-fledged writing system characterized by complex combinations of symbolic imagery (logograms) and phonetic markings (syllabograms) with a fully developed syntactical structure. Like the Mesopotamian and Chinese systems, Mayan writing was primarily a tool of literate elites—the vast majority of the populace continued to communicate with each other via the spoken word.

The linguist Walter J. Ong has charted a pattern of social transformation that appears to recur across human cultures that have made the transition to literacy. Writing invariably emerges at first to support commerce. This phase marks a period of "craft literacy," during which "writing is a trade practiced by craftsmen, whom others hire to write a letter or document as they might hire a stonemason to build a house, or a shipwright to build a boat."[6] Ong suggests that the transition from oral to literate cultures boils down to a few core distinctions: oral traditions are additive rather than subordinative, aggregative rather than analytic, empathetic and participatory rather than objectively distanced, and situational rather than abstract. The written word also enables individuals to transcend the limits of subjectivity. As Ong puts it, "Abstractly sequential, classificatory, explanatory examination of phenomena or of stated truths is impossible without writing and reading."[7] "To say writing is artificial is not to condemn it but to praise it," writes Ong. "Like other artificial creations and indeed more than any other, it is utterly invaluable and indeed essential for the realization of fuller, interior, human potentials." This, then, is the paradox of literacy: It enables us to externalize our experiences and share those experiences with utter strangers, while simultaneously fostering deeper and deeper levels of introspection. "Technologies are not mere exterior aids but also interior transformations of consciousness, and never more than when they affect the word." And although technology inevitably relies on artifice, "artificiality is natural to human beings."[8]

The newly literate, clay-based information economy would dominate the ancient Near East for the next 2,500 years, as networks of trade spread to neighboring regions that in turn created their own local scripts. The yeomen Mediterranean scribes, equipped with their styli and lumps of clay, provided a bedrock layer of trust that supported a growing network of commercial activity. And even though the oral and literate cultures coexisted, as they do today, it was the technologies of literacy that ultimately fostered the transition toward increasingly more complex societal and governmental systems. Just as beads and bone totems had enabled Ice Age people to expand their social circles, so writing served a similar purpose, enabling otherwise unknown parties to forge reliable social bonds cemented by new forms of documents like contracts, laws, and religious rites. The new technology thus facilitated another degree of social abstraction, enabling people to coalesce into farther-flung social networks. From these networks, new kinds of hierarchies would soon emerge.

The Knowledge Bureau

As literary craftsmen began to accumulate power over the conduct of day-to-day transactions, a new kind of collective entity began to form around them. Just as ancient bacteria had coalesced into complex organisms, so the ancient scribes slowly came together into larger social organisms: institutions.

Although writing had originated as a tool for helping individuals buy and sell, its use gradually expanded to encompass larger-scale governmental and religious functions. The economic power of writing afforded scribes a social status second only to the king, increasingly concentrating power in the hands of a literate bureaucratic elite. The escalating complexity of these relationships, coupled with the need for third parties capable of providing both authority and trust, gave rise to multiperson knowledge bureaus predicated on the power of the written word. As human settlements grew, the stewardship of writing slowly passed from the hands of individual craftspeople to fall under the purview of new organizational entities: institutions. The first government bureaucracies, religious temples, and educational institutions began to take shape.

As writing gave rise to increasingly complex institutions, writing itself became more complex. At the Ebla site in Syria, archaeologists found a literary treasure trove of two thousand tablets dating as far back as 2300 BC. Most contain records of routine transactions involving textiles, olive oil, cereals, and other staple products. The Ebla site also contains another set of tablets, set apart from the rest in a neat pile atop the other manuscripts. These tablets contain different kinds of lists that serve no apparent transactional purpose: lists of birds and fish, geographical locations, the names of professions. One tablet consists solely of a list of the other tablets (though not in any discernible order). Another such "meta"-document appears at the ancient Hittite site of Hattusas, near Ankara, Turkey, where archaeologists discovered what appears to be the first known bibliographic record: a list of other documents that includes title pages and rudimentary colophons containing brief descriptions of each tablet. For example,

"Eighth tablet of the Dupaduparsa Festival, words of Silalluhi and Kuwatalla, the temple priestess. Written by the hand of Lu, son of Nugissar, in the presence of Anuwanza, the overseer."[9]

In addition to information about the author and subject of the document, each colophon included a number, indicating that the tablets were stacked in a predictable sequence. The colophon also allowed for a brief set of keywords to pro-vide readers with a glimpse of the document's contents, a kind of primitive abstract to help prospective readers limn the contents of a tablet before invest-ing the time to read it. The Hittites had mastered all the basic functions of a li-brary catalog, capturing information about authors and subjects in a digestible form and adding call numbers to help users retrieve the document they wanted. Melvil Dewey would have been proud.[10]

List making has always constituted an important form of writing. Long before anyone started producing the familiar narrative documents that we associate with writing today, the first true writers were list makers. The first written lists emerged as a growing tide of transactions produced more and more documents, forcing people to develop new mechanisms for organizing the growing sprawl of data: to-kens, then lists, then lists of lists, then bibliographical systems for organizing en-tire collections. By 3000 BC, the scope of written knowledge had grown to the point where Sumerian temple libraries contained commercial records, grammars, mathematics texts, books on medicine and astrology, and religious scriptures. Also fueling the Sumerian writing boom was the emerging institution of the temple. Ancient temples functioned as more than just purveyors of ritual; they were also the first national banks, lending out money at interest to individuals and small businesses. The Sumerian savings and loan managed an enormous volume of transactions, each recorded on a separate clay record, which in turn found its way into periodic "roll-up" reports issued on a weekly, monthly, and annual basis. Indeed, although we might think of information technology as a newish field, in fact information technologist may rank among the world's oldest professions.

Invoices and accounting records paved the way for records of taxes and trib-utes, property records, deeds, and property transactions, with graphic symbols confirming their legality. These early records, written on clay tablets or papyrus or engraved in bronze or copper, formed the foundation of government archives that would eventually grow to include laws, decrees, property records, contracts, trea-ties, and chronicles of events involving the state itself: the outcome of battles, the succession of monarchs, and other chronologies. Slowly, these chronological rec-ords took on the trappings of literature. "Since records of military conquests and biographies of kings often included as much fiction as fact, they added an element of literature to an otherwise staid collection."[11] Over time, the great stream of oral tradition found its way into print.

As the scope of recorded information expanded, scribes began dividing into professional specialties with increasingly narrow domains of knowledge: the *dub-sar kengira* specialized in the classics, the *dubsar nishid* plied mathematics, the *dubsar ashaga* labored at geometry. In the temples, scribes began to write down old spells and legends, slowly. As academic specialties began to coalesce, temples began to establish formal writing education programs to ensure the continuity of skills from generation to generation. As growing numbers of scribes mastered the literary arts, they began to produce more varied texts on, for example, astronomy, prophesies, and scientific observations.[12]

The prolific Sumerian scribes created such a bulwark of recorded knowledge that by the time Sumerian civilization fell into decline, it had created a written legacy that would long outlast its own history. Eventually, a new Semitic power rose in Babylon, displacing the Sumerians. Although the Babylonians spoke their own language, they preserved the written record of Sumerian religious texts, whose ancient pedigrees lent them an air of authority not unlike the veneration that more recent cultures have ascribed to the Bible, Koran, or Torah. As the Babylonians displaced the Sumerians, they placed a great emphasis on preserving the old civilization's textual legacy in much the same way that European nations went to great pains to preserve the ancient legacies of Greek and Latin literature. Like learned Europeans during the Middle Ages, the Babylonians established a bilingual standard: with a spoken vernacular for the common people and an exalted written text, Sumerian, for preserving ancient holy truths.[13] Why did the Babylonians invest so much effort in maintaining old texts in the language of an older empire? The historian A. C. Moorhouse speculates that for the Babylonians, "there was a deep bond between producing documents and maintaining political legitimacy." Recognizing the essential relationship between the written word and political power helped cement the authority of the state over the older, preliterate, symbolically minded tribal cultures.

Indeed, the relationship between written language and political legitimacy stretches deep into antiquity. Just as the earliest literate cultures had invented fables to explain the spellbinding power of the written word, later civilizations would invoke mythologies to assert the relationship between the written word and the political authority of the state. Ancient Romans attributed the prosperity of their empire in part to the purchase of three divine books by the ancient king Tarquin. According to the story, Tarquin bought the volumes from the prophetess Sibyl only after spurning her original offer of nine books, six of which she proceeded to burn out of spite. Realizing his mistake, Tarquin quickly came to his senses and snapped up the remaining volumes. Those books would later occupy a place of honor in the Roman forum, providing a tangible bridge from the mythic world to the present, until they were finally destroyed along with the empire during the great sieges. The Assyrians assigned a similar mythological significance to the power of writing in their tale of Zu, a lesser god who steals a divine tablet from the ruling god Enlil and brings it to Assyria. The tablet is said to reveal the fate of the gods, thus granting the Assyrian kingdom a measure of power over the gods themselves.

Throughout the ancient world, writing played a crucial role in the expansion of empires. In the seventh century BC, King Ashurbanipal took control of Mesopotamia, establishing the great Assyrian Empire. As one of his first acts, he commanded his scribe Shadunu to collect every available written artifact in the kingdom. "No one shall withhold tablets from you," the king decreed, "and if you see any tablets and ritual texts about which I have not written to you, and they are suitable for my palace, select them, collect them and send them to me." Executing the king's decree, Shadunu traveled the countryside impounding every written tablet he could find. He confiscated every single document from every temple and private home in the kingdom. Eventually, he returned with a vast collection of poems, proverbs, hymns, fables, omens, horoscopes, incantations, prayers, and more than five hundred drug recipes.[14] From this act of imperial confiscation came the world's first library. It would not be the last time a ruler recognized the political power of libraries.

In addition to a vast array of factual records, King Ashurbanipal's collection included a literary component: the now-famous legend of Gilgamesh, twelve tablets of the old Babylonian account of the great flood, and numerous translations of older Sumerian stories. Of the twenty-five thousand tablets later excavated at the vast Royal Library at Nineveh, more than two-thirds were created in response to orders from the king.

Just as Ptolemy would later build his library at Alexandria by confiscating books from incoming ships, Ashurbanipal built his library through coercion. And lest there be any doubt about who owned the books now, the king ordered that every book in the kingdom carry this royal inscription: "Palace of Ashurbanipal, king of the world . . . who has possessed himself of a clear eye and the choice art of tablet-writing, such as none among the kings, my predecessors, had acquired. The wisdom of Nabu, the ruled line, all that there is, have I inscribed upon tablets, checking and revising it, and that I might see and read them, have I placed them within my palace."[15]

Some manuscripts he further recommended to posterity, "for the sake of distant days." The books also included a warning that anyone who stole or vandalized a book would be subject to a curse "terrible and merciless as long as he lives." The king also instituted rules requiring a high degree of bibliographic control: every text would be copied in a consistent format, bearing the name of the scribe and the name of the king. The library also bore several hallmarks that would become characteristic of subsequent early libraries: forming part of a temple dedicated to the worship of a deity, following stated acquisitions guidelines, and employing a dedicated staff conversant in multiple languages. Of the organization of the library we know almost nothing, but we do know that it proved instrumental in supporting the expansion of the Assyrian Empire. Armed with a vast arsenal of recorded information, the empire could assert its dominance over the region, its power secured with a bulwark of knowledge that gave it an enormous strategic advantage over potential rivals. Having secured a storehouse of recorded knowledge about governance, military strategy, weapon making, agriculture, mathematics, and other tactical know-how, the empire stood poised to dominate the region.

The earliest libraries were first and foremost vessels of political power, consolidating the accumulated intellectual capital of the early nation-states and providing a durable link with the past by invoking religious authority and asserting a relationship to the gods. The gods, by extension, protected the library, and the genealogical relationships of the gods, echoing old folk taxonomies, found a new manifestation in the nested hierarchies of state institutions.

The first libraries existed primarily to support these growing imperial hierarchies. Indeed, every great empire has boasted of a great library. In China, the earliest known library dates to 1400 BC. In Egypt, Rameses II established a sacred library at Thebes in 1225 BC. The first Indian manuscript collections date as far back as 1000 BC. Every great imperial civilization seems to have progressed along a markedly similar (though far from identical) path: agricultural settlements developed a commercial facility for writing, enabling them to make the transition from tribal societies to nation-states. As some of those nation-states grew into empires, they began producing more varied forms of literature that were eventually gathered into libraries.

The fates of those libraries would prove no less turbulent than the empires that built them. Indeed, the advent of literacy and book making has invariably been accompanied by violence and political turmoil. When Emperor Shi Huangdi consolidated power over the Chinese Empire in 213 BC, he promptly ordered an imperial biblioclasm, commanding the destruction of every book in the kingdom. Soldiers demolished the old royal library, a priceless trove of early Confucian and Taoist texts known as the Heavenly Archives (whose most famous curator was Lao Tzu). After clearing the brush of the prior regime's intellectual legacy, the emperor created a new library, complete with a new classification system to reflect the new imperial order. That system eventually coalesced into the great ancient Chinese library classification system known as the Seven Epitomes. The scheme divided the library's holdings into six top-level categories: military, science, technology, medicine, poetry, and philosophy. The epistemological hierarchy also manifested itself in the political structure of the new empire. As a ministerial administration took shape to support the new imperial priorities, the hierarchy of the library became the hierarchy of the state. But this new classification, like the one it supplanted, would last no longer than the empire that supported it. When the Wei dynasty later took power, the new imperial librarian Cheng Moe established yet another new classification scheme to replace the old one, reflecting a new set of imperial priorities, boiling everything down to four categories: classics, philosophy, history, and literature.

Centuries later, when the Spanish conquistador Hernán Cortés arrived in sixteenth-century Mexico, he discovered a civilization that had mastered the art of book making a full thousand years earlier. The Aztec libraries contained an enormous collection of manuscripts about history, myth, biographies, herbalism, mathematics, and all kinds of local lore. When the conquistadors arrived, Bishop Diego de Landa ordered them to burn every one of the bark-clothed books because they "contained nothing but superstitions and falsehoods of the Devil."[16] Although it may be tempting to bemoan the wanton destruction of an Indigenous culture at the hands of a European imperialist power, the Spanish were not the first invading force to raze an Aztec library. Just a century before the Spanish arrived, a conquering Aztec ruler named Itzcóatl burned the previous royal library to the ground, clearing the intellectual brush for a new Aztec history written in his own mold.

The violent history of libraries is a mirror of empire building: hierarchical systems emerging from violent political upheavals, only to collapse, disintegrate, and give rise to new emergent systems. Ever since the first temple libraries emerged in Sumer around 3000 BC, libraries have flourished and fell with the empires that supported them. From ancient Sumer to India to China to the Aztec kingdom, the same pattern manifested again and again: first came literacy, then the nation-state, the empire, and ultimately the intellectual apotheosis of the empire, the library. When empires fall, they usually take their libraries with them.

The greatest library of the ancient world, of course, was the legendary library at Alexandria. Today, Alexandria enjoys an almost mythic status as the archetypal temple of ancient wisdom. Why does Alexandria, alone among all the great libraries of antiquity, occupy such a special place in the popular imagination? Alexandria was more than just a large collection of books; it was the apotheosis of a whole new way of thinking that departed radically from the old imperial information systems of the preclassical era. To understand why Alexandria still matters, we need to understand a thing or two about the Greeks.

From Mythology to Ontology

Sometime after 2000 BC, troops of mounted warriors swooped down from the Caucasus Mountains, through the Balkans, and into the valleys of the Greek peninsula. They terrorized the Indigenous population of farmers, ultimately subduing them and establishing a series of new kingdoms throughout the region. During the so-called Mycenaean Age (approximately 1600–1200 BC), a succession of small monarchies emerged that looked a lot like other regional kingdoms throughout the Near East. Like the Sumerians and Babylonians before them, the Minoans and Mycenaeans established a caste system predicated in part on the legitimating power of the written word. A ruling class of kings, attended by literate scribes and priests, established royal institutions, hoarded wealth, and wielded power over the illiterate populace of farmers and soldiers. Once again, it seemed, literacy was emerging along a familiar trajectory, with the technology of the written word serving to prop up ruling institutional hierarchies. But this time, things would turn out differently.

The first Greek civilization came crashing to an end around 1200 BC, through a series of political disruptions that would be immortalized in Homer's *Odyssey*. The first Greek kingdoms collapsed, and the region descended into an ancient equivalent of the Dark Ages. The art of literacy deteriorated and then vanished. State institutions disintegrated, and Greek culture vanished into a historical black hole, leaving almost no written trace for more than two centuries. As the first Greek kingdoms collapsed, people reverted to earlier tribal ways, re-embracing their old oral traditions. The disintegration of institutional hierarchies created a fallow ground for ancient social networks to reemerge.

In this renewed oral culture, ancient myths took on new vitality. As we saw in chapter 2, these mythologies were more than just popular stories; they reflected a deep-seated disposition to understand the world in terms of family relationships that bear a close resemblance to the structures of folk taxonomies. The tales of Greek mythology originated from earlier traditions in Sumer, Babylon, Ugarit, and Boğazköy, which in turn derived from even older tribal myths. These overlapping streams of narrative myth extended like a vast cultural watershed across the ancient Near East. In the structure of these ancient stories, we can find the echoes of old taxonomic hierarchies. Hesiod's *Theogony*, the canonical version of the Greek myths, portrays a family tree of the gods that runs about five to six levels deep, just like a folk taxonomy, with each god representing a particular element of the natural world. Zeus is the god of thunder, Poseidon the sea, Hades the underworld, and so on. The relationships between the gods describe allegories and metonymies encoded in a structure of "family" relationships.

Figure 11 shows a partial glimpse of how the family relationships between the gods unfold. Chaos, the primordial abyss, gives birth to Gaia, the earth, who in turn gives birth to Cronos (the harvest) who begets the family of gods most familiar to humans: Hestia, Zeus, Hera, Hades, and Poseidon. Zeus and Hera, the archetypal king and queen, give birth to Athena the goddess of civilization, Ares the god of war, and Hephaestus the god of technology. Ares gives birth to Phobos

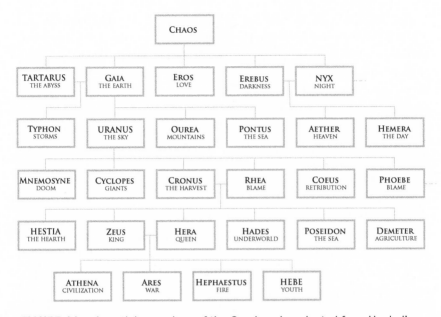

FIGURE 11.   A partial genealogy of the Greek gods, adapted from Hesiod's *Theogony*. © Alex Wright.

the god of fear and Harmonia the goddess of harmony. The whole system functions as a kind of cultural encyclopedia, cross-referenced using the familiar structure of family genealogy. The stories of Zeus and the gods also narrate metaphorical relationships between natural phenomena. For example, Okeanos and Tethys give birth to rivers and streams; Hyperion and Thela to the sun, moon, and dawn; and so forth. The familiar myth of the Muses—the divine beings said to inspire poets and artists—represents more than just a superstitious belief in the possibility of divine inspiration. "The Muses' genealogy is transparent allegory," writes the classicist Robert Lamberton. "Offspring of power and memory, they embody the prestige and authority of the highest of the gods. Memory, in this context, is not the capacity eroded by Alzheimer's or obliterated by Korsakov's syndrome but rather a fine-tuned programming in a vast range of cultural skills and information."[17] The *Theogony* thus serves as a taxonomic prototype for the structure of Greek thought.

Although it would be overly simplistic to suggest that the rich tapestry of Greek mythology can be reduced to a singular taxonomy, nonetheless the basic structure of these divine genealogies seems to echo the implicit structures of ancient folk taxonomies. The most popular gods, the Olympians, appear roughly in the middle of the hierarchy, in a position corresponding to the genus, the "psychologically primary" category in folk taxonomies (see chapter 2).

The structure of the gods' families holds glimpses of older oral traditions that predated the first era of Greek kings, traditions that spread through the pure network of oral traditions, unfettered by priestly bureaucracies. "From the earliest glimpses we have into Greek tradition, the poets and visual artists, not the priests, were the bearers of the traditions about the divine," writes Lamberton. "The contrast with the priest-ridden cultures of the East has often been drawn but cannot be too much emphasized. The Greeks had no Moses, and their first theologians were entertainers—Homer and Hesiod."[18] In the years following the disappearance of the old state hierarchies during the Greek Dark Ages, poets strengthened their command of the old stories and, by extension, the transmission of cultural tradition.

By the time Greek civilization reemerged from its writing-less silence in the ninth century BC, the Greeks were living in a new world. Poetry predominated as their primary information system. Soon it would form the cultural bulwark of their new literate civilization. Before the Greeks could reestablish stable political systems, however, they first had to rediscover the art of writing. Whereas the earlier Greek civilization had used a form of writing derived from the imperial Babylonian script, this time the Greeks borrowed a new script from their trading partners, the Phoenicians. The Phoenicians used a phonetic writing system, with letters signifying sounds rather than ideas. The Greeks adapted this alphabet to their native oral language, in the process hitting on a radically simplified style of writing. Their new alphabet contained just twenty-four symbols. It was easy to learn. In the earlier Greek civilization, literacy had required many years of training and imperial support; it was only for the elites. Now almost anyone could learn to read and write. This time, the Greeks would not succumb to the old imperial "tyranny of the book."[19]

The Greeks took up the new form of alphabetic writing with enthusiasm. Whereas the old empires had used writing largely for affairs of state—government archives, administrative records, hagiographies of their rulers—the newly literate Greeks put writing to more populist ends. Emerging from a period of reengagement with their older oral heritage, mythologies, and folk traditions, the Greeks started writing down their old stories. By the time of the great civic festivals of Athens in the sixth century BC, readers were clamoring for standardized versions of their old myths. This period saw a great shift in the nature of storytelling, from an era when poems enjoyed a fluid structure in the ever-shifting versions of oral storytellers to a more rigid and deterministic form, as poems became "frozen and shackled to the interests of political power," as Lamberton puts it. Although the stories of the Greek myths contained embedded hierarchies of family relationships between the gods, those hierarchical narratives fostered the bonds of social networks. When the stories became ensconced in writing, however, new forms of hierarchy emerged. "Institutions demand—indeed, impose—stability and are able to harness the power of even such unlikely sources as poetry."[20] Although the notion of poems as political tools may strike us as implausible today, the poems of ancient Greece—like the *Odyssey*, or Hesiod's mythology—went straight to the heart of questions of political legitimacy. The emerging Greek political institutions had a clear vested interest in nailing down standard versions of these texts as safeguards of their own power. These old tales, preserved through the Greek Dark Ages through the aegis of social networks, became important tools in establishing new political hierarchies.

As the old myths became ensconced in writing and embedded in a growing bureaucratic structure, the Greeks began to approach their old stories in a new light. At first, most of these stories had fallen into the category of what the psychologist Merlin Donald calls mythic or integrative thought in which cultural knowledge is synthesized in the form of densely layered mythologies. Mythic thought, though it may be expressed in writing, harkens back to the oral traditions of tribal cultures. The Greeks, however, took their understanding a step further. They were, as far as we know, the first culture to make the leap from mythic to "theoretic" thought—that is, the ability to reflect on the process of thought itself. Writing gave Greeks the chance to approach their old myths with a new reflexive stance, cultivating a degree of objectivity that enabled them to compare, assess, and excavate new layers of meaning from the texts. By writing down their old stories, they made the inner logics of these systems visible. Once the text became externalized, it could be subjected to analytical thought, reworked, and even improved on. This process of textualization corresponds to what Donald calls the demythologization of culture. "The first step in any new area of theory development is always antimythic," he writes. "Things and events must be stripped of their previous mythic significances before they can be subjected to what we call 'objective' theoretic analysis. In fact, the meaning of 'objectivity' is precisely this: a process of demythologization." For example, astronomy constitutes the demythologization of astrology, chemistry the demythologization of alchemy, anatomy constitutes the demystification of the human body, and so on. "Before nature could be classified and placed into a theoretical framework, it too had to be demythologized," writes Donald. "Nothing illustrates the transition from mythic to theoretic culture better than this agonizing process of demythologization, which is still going on, thousands of years after it began. The switch from a predominantly narrative mode of thought to a predominantly analytic or theoretic mode apparently requires a wrenching cultural transformation."[21]

This second coming of alphabetic thought left some Greek thinkers with a conflicted relationship to the emerging information technology of the written word. In *Phaedrus*, Plato recounts the story of the Egyptian god Theuth, the discoverer of writing, the "speech of the gods." When he presented his discovery to the Egyptian god king Thamus, presumably expecting accolades for his breakthrough, the king instead gives him a sharp rebuke. "This discovery of yours will create forgetfulness in men's souls," says the king, "because they will not use their memories; they will trust to the external written characters and not remember of themselves." The king goes on to decry the art of writing as "an aid not to memory, but to reminiscence," warning that readers will become "hearers of many things and will have learned nothing . . . having the show of wisdom without the reality."[22]

This story has often been misinterpreted as Plato's condemnation of writing. In fact, Plato was a great lover of books. He maintained one of the largest private libraries in Athens. The tale is actually a veiled attack on the Sophists, who devoted themselves to the enthusiastic production of knowledge without, in Plato's view, a proper reverence for reading. The Sophists' disposition toward profligate penmanship prompted Aristophanes to deride the ancient Greek capital as "Athens full of scribes."

Plato's story also contains the central paradox of writing: Plato surely knew the story could not possibly have survived without being written down. The story seems to recognize the threat posed by literate culture to earlier oral traditions. Yet the interpolation of oral and literate cultures was the essential factor in the Greek contribution to human knowledge.

Poised at the cusp of the transition from oral to literate traditions, the tales of Greek mythology provide a perfect boundary object for analyzing the effects of this profound cultural shift. They give us a revealing glimpse into the mythic world of the oral traditions that preceded the reemergence of Greek literacy, showing how mythologies rooted in ancient wisdom traditions provided the cultural framework for the great Greek experiment with decoding the process of "knowing" itself. "From the first, Greek thinkers were employing external memory devices to their fullest effect, in a way that was totally new," writes Donald. "In modern culture, narrative thought is dominant in the literary arts, while analytic thought predominates in science, law, and government. The narrative, or mythic, dimension of modern culture has been expressed in print, but it is well to keep in mind that in its inception, mythic thought did not depend upon print or visual symbolism; it was an extension, in its basic form, of the oral narrative."[23]

Analytic thought gives rise to "formal arguments, systematic taxonomies, induction, deduction, verification, differentiation, quantification, idealization, and formal methods of measurement"—systematized ways of understanding the world that are all absent from the world of mythic thought.[24] "The natural tendency of spoken thought . . . is toward fairly loose narration of events, metaphoric fantasy, and storytelling. Fine-grained analysis of the thought process itself is difficult, because the memory trace of an oral narration is so ephemeral."[25] The new art of "thinking about thinking" opened the door to introspection—formalized ideas and conjectures, unfinished thoughts, and dialogue. The process of demythologization not only affects the forms of human thought but also influences macrostructures of social organization. Whereas oral traditions lend themselves naturally to fluid self-organizing social networks, written analytic thought creates a different imperative, fostering the kind of linear thinking that seems to beget organizational hierarchies. Donald speculates that "the governing cognitive structures of the most recent human cultures must be very different from those of simple mythic cultures. They exist mostly outside of the individual mind, in external symbolic memory representations, which are dependent upon visuographic invention, and they culminate in governing theories."[26] In other words, the shift from mythological to analytic culture played out in the evolving structure of the state itself.

The journey from mythic to analytic thought was hardly a smooth one. Socrates was put to death for lack of reverence toward the gods, and Aristotle later faced the same accusation. But slowly, inexorably, the old mythologies began to make way for a more systematic way of understanding. The process of demythologization that started in ancient Greece continues to this day. Although we may live in a highly analytic, systematized culture, oral myths continue to exert a strong pull. Orality has hardly disappeared; indeed, with the rise of electronic media, it may be reemerging. Ong has suggested that we are witnessing the birth of a "new orality" in the form of electronic communications. Just as the Greeks of the classical Dark Ages saw new forms of thought emerge when oral and literate cultures began to interpolate each other, we may be witnessing a similar period of cultural renegotiation today, a process whose outcome we probably cannot yet predict.

If the Greek journey from mythic to analytic thought tells us anything, however, it may point to the sheer durability of mythological thinking. Today, most of us still know more about Greek mythology than we do about classical logic. Elementary school students can still tell you stories about Zeus, Hera, and Apollo, while few of them could tell you much of anything about Aristophanes. The sheer longevity of these stories, even in our technologically advanced society that ought to be well along the path to demythologization by now, suggests that the enduring resonance of myths may stem from a deeper affinity in the human psyche. Perhaps the ancient structures of folk taxonomies equip us with a subtle longing to understand the world in terms of gods and their families.

From Aristotle to Alexandria

As Greek civilization spread throughout the ancient Near East, so too did its books. Long before Alexander established his great eponymous library in Egypt, the Greeks had established great halls of learning at Antioch and Pergamon, a library said to contain more than two hundred thousand scrolls. Athens was the bibliographical center of the Greek world, playing host to a bustling book trade that fueled the rise of personal libraries among a growing class of literate citizens. As public literacy flourished during the age of Sophocles, Aristophanes, Thucydides, Euripides, Aeschylus, and the other great secular writers of ancient Greece, private citizens began to amass growing book collections. By the fourth century BC, a typical Athenian library might contain works by Homer and Hesiod, tragedies, histories, works of poetry and philosophy, and even cookbooks. The largest and most famous of Athens's personal libraries belonged to Aristotle, who not only amassed great numbers of books but also developed a keen fascination with the question of how to organize his collection. As the scholar Strabo wrote, Aristotle "was the first to have put together a collection of books and to have taught the kings in Egypt how to arrange a library."[27]

Aristotle's interest in organizing books grew out of his passion for mapping the structure of human thought. He devoted enormous energy to questions surrounding the organization of knowledge, postulating theories of logic and taxonomic models that would reverberate for millennia. Aristotle's work on categorization would provide the foundation for a contribution to the heritage of Western thought too vast to be addressed in remotely adequate detail here. Suffice it to say that students of philosophy and the social sciences have spent thousands of years pondering Aristotle's categories.

Aristotle took a special interest in the organization of knowledge about the natural world, devoting a great deal of his work to the search for logical order in the world of visible phenomena. He championed deductive reasoning, or the process of drawing axiomatic theories from empirical observation. He grounded his work in an exhaustive program of scientific observation, gathering vast amounts of firsthand data about plants and animals, which he then labored to situate them in a larger categorical framework. Although Aristotle's writings about plants have been lost altogether, his works on the animal kingdom constitute a seminal contribution to biology. He proposed a comprehensive taxonomy of animal life that distinguished mammals from nonmammals, vertebrates from invertebrates, even recognizing that whales and dolphins differed from fish (although he failed to recognize them as mammals). He also introduced the convention of binomial naming—that is, describing a species by its genus (e.g., felis) and species (e.g., catus). Reflecting the dominant view of his day, Aristotle mistakenly believed that animal species did not change over time but rather that they represented instances of fixed, immutable ideal forms stretching back into time immemorial (Darwinian evolution would have been a thoroughly alien concept to the ancient Greeks).

TABLE 2

VERTEBRATES	INVERTEBRATES
Quadrupeds (or mammals)	Cephalopods
Birds	Crustaceans
Reptiles	Insects
Fish	Shelled animals
Whales	Zoophytes (or "plant-animals")

Although Aristotle is often credited as the first great taxonomist, his system almost certainly relied on earlier folk taxonomies that had persisted for thousands of years in earlier oral traditions. Although his naturalist writings would have a lasting impact on the subsequent course of biology, he receives perhaps a little too much credit as the world's first taxonomist. Aristotle did not invent the concept of genera and species; he simply adapted and codified the ancient oral traditions of folk taxonomies. His great contribution was to make the old implicit structures explicit and therefore available for inspection and improvement. In so doing, he extracted an analytic system of thought from the remnants of oral traditions shrouded in myth. If we understand Aristotle's categories not just as the insights of a great thinker but also as the fruition of ancient folk taxonomies long in the making, then perhaps we can also speculate as to why his system (flawed though it was) persisted for thousands of years. By invoking the structure of folk taxonomies, Aristotle's categories may have resonated with a cognitive rule set lodged deep within the prehistoric mind. Although Aristotle's landmark writing on the natural world owed a great deal to the heritage of folk taxonomies, it also signaled the beginning of the end of the old folkways and the inauguration of a new systematic approach to scholarship.

Aristotle did a great deal more than simply categorize plants and animals, of course. Building on his research into the natural world, he went on to postulate a metaphysical worldview that drew the crucial distinction between matter and form, or what he called primary and secondary substances. Primary substances referred to the individual instances of particular phenomenon; secondary substances referred to their ideal forms, the disembodied templates of human thought.

> Substance, in the truest and primary and most definite sense of the word, is that which is neither predicable of a subject nor present in a subject; for instance, the individual man or horse. But in a secondary sense those things are called substances within which, as *species*, the primary substances are included; also those which, as *genera*, include the species. For instance, the individual man is included in the species

"man," and the genus to which the species belongs is "animal"; these, therefore that is to say, the species "man" and the genus "animal"—are termed secondary substances.[28]

By recognizing the distinction between form and matter, Aristotle believed the human mind could aspire to penetrate higher realms of truth. In his *Categories*, Aristotle enumerated a comprehensive set of classes and subclasses that he believed could reveal the structure of nature's creation. He believed that any phenomenon in the world could be described in terms of a particular set of characteristics: substance, quantity, quality, relation, place, time, position, state, action, and passion. He further believed that the relationships between phenomena could be explained in terms of a web of particular semantic relationships: causality, equivalence, identity, similarity, family, inside of, bigger than, and earlier than. By weaving the universe of forms together with these relational threads, Aristotle conceived of human knowledge as a vast interconnected tree that extended even beyond the observable world, into incorporeal realms. His famous great chain of being classified all life into a grand cosmic hierarchy, from the lowest realm of minerals to the highest realm of the gods. Unlike the traditional folk taxonomies that concerned themselves with the observable world of plants and animals, Aristotle's taxonomy stretched into the otherworldly domains of gods, angels, and demons:

Aristotle's Great Chain of Being:

Gods
Angels
Demons
Man
Animals
Plants
Minerals

Aristotle's categories would cast a long shadow over the subsequent development of the Western intellectual tradition, influencing the trajectory of European thought, especially during the Middle Ages, when scholars rediscovered Aristotle and elevated his writings to the level of canonical truth second only to the Bible and the church fathers. Aristotle's taxonomic conventions would persist for more than two thousand years until Carl Linnaeus embraced and improved on the Aristotelian system to create the familiar modern Linnaean taxonomy still in use today (see chapter 9). Aristotle's work on categorizing the phenomenal world reflects his deeply systematic worldview, his conviction that the world could be described in terms of ideal forms. His quest for logical order led to his conviction that the world was ultimately knowable. That conviction not only drove Aristotle's personal ambitions but also served—along with his personal library—as the intellectual foundation for the greatest library of the ancient world.

The Universal Library

The Great Library at Alexandria was the first library with a truly comprehensive ambition to gather all the world's knowledge under one roof. Established around 300 BC, the library marked an achievement of vast intellectual proportions, ultimately growing to house more than seven hundred thousand items. Unlike the royal libraries that preceded it, the Alexandrian library was open to the public. By some accounts, the library literally began with Aristotle's own personal collection (other accounts insist that the library only took its conceptual inspiration from Aristotle and that his book collection was buried in Athens after his death, later to be dug up and sold to book collectors). Whether the library contained Aristotle's actual books, there is no question that it embodied his ideals. The structure of the library followed Aristotle's division of knowledge into observational and deductive sciences. The library was also built specifically with Aristotle's peripatetic ideal of scholarship in mind, with wide colonnades and open spaces to encourage scholars to stroll around and converse about topics of scholarly interest.[29] We can imagine scholars strolling the grounds, papyrus rolls in hand, debating the politics of the day or perhaps a fine philosophical point. In keeping with the great teacher's insistence on open inquiry, the library encouraged "walking around," both physically and metaphorically. To Socrates, Plato, and Aristotle, intellectual freedom was a paramount civic value (they would surely have bristled at the scholarly territorialism and narrowcast specialization of the modern academy), and the architecture of the library reflected that point of view.

Today, Alexandria occupies a special place in the popular imagination as a symbol of the lost era of classical scholarship and as the site of one of history's great biblioclasms, the place where the great tide of classical learning rose, crested, and finally crashed into oblivion. But the real story of Alexandria turns out to be considerably less tidy than the myth. Its birth was less idealistic, its heyday less peaceful, and its demise less cataclysmic than the popular fable suggests.

When Ptolemy I established the royal library, Alexandria was a new city, populated by military people, traders, and assorted hangers-on. As an intellectual center, it could scarcely hold a candle to Athens. But the city's first ruler, Ptolemy, fancied himself a scholar, having composed a detailed history of Alexander's conquests and studied geometry with no less a mathematician than Euclid. Determined to attract scholars to his young city, Ptolemy offered enormous incentives for learned men to come to his new metropolis at the mouth of the Nile. He

FIGURE 12.   Artist's rendering of the Great Library at Alexandria, from *Histoire générale des peuples*, 1880.

invited poets, writers, and scientists to live in the city and work at its famous museum, offering them room and board in the royal palace and a generous tax-free salary, as well as the guarantee of lifetime employment. As a final inducement, he built the world's first great library (actually called a museum), whose ambitions were nothing short of universal. By the time of Ptolemy's death, the collection had grown to two hundred thousand books. By the time Julius Caesar arrived in 47 BC, the library had swelled to seven hundred thousand volumes. The library's signal achievement was the sheer magnitude of its collection, orders of magnitude larger than any library before it and larger than any library that would follow for more than a thousand years.

The library's stated acquisition policy was simple: acquire everything by whatever means necessary. In contrast to the popular image of the library at Alexandria as a halcyon enclave of scholarship, in truth the library was born out of conquest, populated through a policy of government-sponsored theft and confiscation and propped up by royal cronyism. The Alexandrian rulers built the great library not just as an act of imperial generosity but also through fiat, confiscation, and occasionally subterfuge. According to one story, Ptolemy III sent to Athens for a collection of tragedies by Aeschylus, Sophocles, and Euripides, giving them a substantial cash deposit along with a promise to copy the manuscripts and send the originals back to Athens. Upon receiving the prized volumes, the king promptly reneged on his promise, keeping the original and sending back a copy instead. As the library acquired new volumes, it stored them in a vast warehouse where they awaited cataloging (library acquisition backlogs are nothing new).

Like other great libraries throughout history, the Alexandrian library mirrored the struggles of the empire that begat it. Its rise and fall rested entirely on the stability of its patron empires, and like all institutional knowledge systems, it ultimately proved no more durable than its imperial patrons. The Greek heyday of self-organized scholarship was over, and the new imperial age after Alexander would give birth to a great institutional consolidation of knowledge. The emperors' intentions were not, in other words, purely arcadian. They recognized the political value of intellectual capital. Ptolemy's adviser Demetrius saw the importance of a library not only in promoting the flourishing of scholarship but also ensuring the newly relocated empire a competitive advantage over neighboring rivals in terms of science, technology, and statecraft. Toward that end, the library implemented history's most aggressive collection development program: confiscating books from private citizens and insisting that all ships harboring at the docks yield whatever books they carried as the price of entry to the port. And although the scholars at the library enjoyed near-total academic freedom, with a wide mandate to roam where their minds would take them, in truth they were a little less free than we might like to imagine. Although scholars were under no obligation to pursue any particular discipline, they were beholden to the patronage of the emperors. As the Alexandrian institution grew, it began to suffer from the curse of many a state educational organ: intellectual conservatism. The residents of the so-called museum were putatively free to pursue their intellectual interests, but royal patronage exerted a subtle and ultimately stultifying influence. None of the Alexandrian scholars, so far as we know, felt quite so free as to question the imperial system of government. Alexandria became a haven for scholarly sycophants or, as Timon of Phlius put it, a "chicken coop of the muses." As the library historian Leslie W. Dunlap writes, "The Museum typified the derivative culture of Alexandria: significant original creations were rare indeed, but here the laurels of Greek civilization were kept green for seven centuries."[30] Indeed, given hundreds of years' worth of munificent imperial support, perhaps the most remarkable thing about the library is how little original scholarship it actually ever produced.

Alexandria's first librarian was a scholar named Zenodotus, who had won fame for his efforts at standardizing the old Homeric texts. A pioneering figure in library science, Zenodotus introduced the first library classification scheme, a rudimentary subject scheme that assigned texts to different rooms based on their subject matter. At first, the library had no catalog; the collection was its own catalog. Curators attached a small tag to the end of each scroll, describing the work's title, author, and subject. Although this elementary bibliographic data constituted the first systematic abstraction of metadata, those data remained physically affixed to the material book. During the library's initial period, this approach worked well enough as long as the collection remained small enough that readers could peruse its contents simply by walking around the room. One of the early librarians, Aristophanes of Byzantium, is said to have read the entire collection himself.

While the library served as, in effect, its own catalog during the early years, eventually the collection reached a size of such magnitude that readers needed a better way to navigate it. The poet and librarian Callimachus was the first to undertake the job of creating a separate catalog of the collection, a comprehensive bibliography known as the *Pinakes*, or "Tables of Persons Eminent in Every Branch of Learning Together with a List of Their Writings." Although he succeeded in cataloging only one-fifth of the entire collection, his catalog was an impressive undertaking, filling no fewer than 120 scrolls and earning Callimachus a place in history as not only Alexandria's greatest poet but also the world's first great bibliographer. Alas, the *Pinakes* failed to survive the library's destruction, but thanks to descriptions by contemporary writers, we know something of its structure. It consisted of a set of tables, broken down in a top-level distinction between poetry and prose, further divided into subcategories. Interestingly, the *Pinakes* was not a classification of works but of authors (apparently, the Alexandrian librarians saw no problem with literary typecasting). From what we know if its structure, the classification broke down as follows:

Within each table, authors' names appeared in alphabetical order, accompanied by a brief biographical description. The catalog was the library itself, with each room representing a broad subject area: verse, prose, literature, science, and so forth. As the Alexandrian collection mushroomed to over five hundred thousand volumes, the task of maintaining the catalog by hand outstripped the ability of the librarians to keep pace. Over time, the library began to institute a two-tiered system that gave preference to major authors, ensuring their historical longevity at the expense of lesser scribes.

The story of Alexandria's demise has long been the subject of competing myths. One story lays the blame for its death at the feet of Julius Caesar, who is said to have burned the library as a defensive ploy against a storming Alexandrian mob. Another story sets the blame seven hundred years later, at the feet of

TABLE 3

POETRY	PROSE
Dramatic poets	Philosophers
Tragedy	Orators
Comedy	Historians
Epic poets	Writers on medicine
Lyric poets	Miscellaneous

the Muslim conqueror Amr, who is said to have followed the advice of a caliph: "Touching the books you mention [at Alexandria], if what is written in them agrees with the Book of God, they are not required; if it disagrees, they are not desired. Destroy them therefore."[31] Other scholars believe the library persisted in some form until about 270 AD, when the emperor Aurelian laid waste to the city while fighting a rabid insurgency. Whatever the circumstances of its demise, we know the library's end came as a result of armed conflict; its fate was inexorably tied to the fate of its empire.

Today, the Great Library at Alexandria stands as an idealized symbol of scholarship and learning. But the tale of its rise and fall also tells an instructive story about the inherent instability of institutional systems. And its burning, though surely a great historical tragedy, marks only one in a long progression of historical biblioclasms.

After Alexandria

Although Alexandria occupies a singular place in the popular western imagination as the iconic library of the ancient world, libraries flourished throughout the Greek and Roman worlds. As Rome began appropriating Greek culture starting in the fourth century BC, Roman citizens populated their private libraries from the great book markets in Athens, Rhodes, and Alexandria. Wealthy citizens collected decorative books to ornament their homes, while less affluent Romans could buy small unadorned books for a few coins. Prominent Roman citizens like Cicero built vast personal libraries. The fashion for book collecting reached such a height that the poet Ausonius satirized the folly of citizens who increasingly purchased books for ornamental purposes:

That thou with Books thy Library hast fill'd
Think'st though thy self learn'd and in Grammar skill'd

Then stor'd with Strings, Lutes, Fiddle-stricks now bought;
Tomrrow thou Musitian may'st be thought.[32]

Shortly before his death in 44 BC, Julius Caesar decreed that a great public library should be built. A statesman named Pollio picked up the imperial charter after Caesar's assassination and started building the first great Roman public library with the support of no lesser public intellectuals than Catullus, Horace, and Virgil. Located near the Roman forum, the library consisted of two distinct collections: one block for Latin works and another for Greek works. This precedent of dividing the library into two parallel collections would persist in subsequent Roman libraries, as generations of Romans built new public libraries modeled on Pollio's original. Centuries later, early Christian libraries would follow this model of bibliographical bicameralism, bifurcating their books into collections for Christian and pagan works.

By the fourth century AD, Rome had no fewer than twenty-eight public libraries. Although none of these libraries came anywhere close to approximating the vast Greek collections at Alexandria and Pergamon, they played a visible role in everyday Roman life. The typical Roman library stood in close proximity to a temple or palace, usually conjoined by rows of columns. Walking inside, the citizen would traverse a front entryway with busts of famous poets and authors, arranged into their separate chambers for Latin and Greek. The larger libraries featured bookcases with numbers corresponding to entries in a master catalog listing. As in Alexandria, each scroll bore a tag identifying its author and title. The bookcases opened out onto reading rooms, where library users could peruse a chosen manuscript; some libraries permitted private borrowing.

Roman libraries also served as intellectual salons, housing literary, social, and political groups. Although only a fraction of Rome's residents would likely have availed themselves of these services, the public mission of the library nonetheless signaled a continuation of the Greek commitment to public scholarship. In many Roman communities, libraries were attached to the public baths. The library system was administered by a *procurator*, with each individual library headed by a librarian with a staff of assistants and copyists who made up the preponderance of the library staff. Librarians enjoyed high esteem during this period. Martial describes Sextus, the librarian of the Palatine, as having "intelligence approaching that of a god."[33]

By the fourth century AD, Roman libraries had spread throughout the empire, from Africa to Britain. Roman scribes distributed their copied manuscripts throughout the conquered territories; these far-flung collections would later prove instrumental in preserving the heritage of classical literature. By the latter days of the Roman Empire, however, public love of learning began to degrade into an all-out embrace of Greek epicureanism. Libraries languished as citizens increasingly devoted their energies to sensory pleasures. Coupled with the pacific influence of Christianity, a love of the good life overwhelmed the old farmer-soldier ethic of the early republic, and the old ideals—including the long tradition of civic bibliophilia—began to diminish. By 378 AD, Ammianus reported an empire whose citizens had relinquished their former love of knowledge in favor of pleasures of the flesh: "In place of the philosophers the singer is called in, and in place of the orator the teacher of stagecraft, and while the libraries are shut up forever like tombs, water-organs are manufactured and lyres as large as carriages, and flutes and huge instruments for gesticulating actors."[34]

In our age of pop music, TV dramas, and Broadway musicals, some of us might resonate with Ammianus's lament for a once-proud literate culture decaying amid a rising popular taste for easy entertainment. Just as the old hierarchies of the Roman Empire started falling into decay, so too did its libraries. By the fifth century, Rome's great public libraries had all but disappeared, either destroyed at the hands of the Vandals or carted away to the new capital in Constantinople. By the sixth century, they were gone. The old hierarchies of the Roman Empire—and its imperial libraries—gave way to a period of intellectual chaos, creating a fallow ground from which new forms of knowledge would soon emerge.

ILLUMINATING THE DARK AGE

A scholar is just a library's way of making another library.

—Daniel C. Dennett

An Irish legend holds that one night in the middle of the sixth century, Saint Columba borrowed a manuscript from his guest Saint Finnian, staying up all night inside the church to make his own copy of the rare text. According to the legend, Columba's fingers shone like candles, lighting up the whole church.[1]

With the collapse of the Roman Empire in the fifth century, Europe lost more than just its imperial government; it also lost thousands of texts—Plato, Aristotle, Cicero, Virgil—the intellectual foundation of the Pax Romana. As the Goths and Vandals swarmed into the city, they burned the libraries and government archives, annihilating the empire's collective intellectual capital. During the so-called Dark Ages that followed, the scorched earth of the old empire gave way to the reemergence of smaller close-knit societies that more closely resembled the earlier tribal chiefdoms of old. Although the institutional template of the old imperial bureaucracies persisted in the form of the growing Roman church, daily life for most Europeans reverted to a more tribal mode of existence. The old hierarchies of the state collapsed, and Europe entered a period in which literacy waned, while older oral traditions and kinship networks reasserted themselves. To invoke the evolutionary metaphor, the Dark Ages were a period of mutation and drift. Across Europe, this seemingly regressive period provided fertile ground for a series of new technologies that would soon begin to take shape. Most of this innovation would happen in a few isolated corners of Christendom, within the cloistered walls of early monasteries.

In the century before the fall of Rome, the unwieldy twelve-foot-long papyrus scrolls of Alexandria had started to cede room on the shelves to a new form of document: the codex book, so named because it originated from attempts to "codify" the Roman law in a format that supported easier information retrieval. The new format boasted a more navigable interface, featuring leafed pages bound between durable hard covers. Not only was the new format more resilient than the carefully wound scrolls that preceded it, but it facilitated a new way of engaging with the text. Scrolls, by their physical nature, demanded linear reading from start to finish. They required a commitment on the part of the reader and resisted attempts to extract individual nuggets of information. But with a codex book, readers could now flip between pages easily to pinpoint any passage in a text. As Michael Hobart and Zachary Schiffman write, "The codex had the potential to transform the manuscript from a cumbersome mnemonic aid to a readily accessible information storehouse."[2]

The new technology of the book ushered in a whole new way of reading: random access. A document no longer had to be read from top to bottom; its pages could be flipped, allowing the reader to move around at will. By letting users move freely from page to page, the new book allowed readers to form their own networks of association within a particular text. Scrolls, on the other hand, encouraged linear reading. The book also had one more considerable advantage over the old scrolls: portability. Although collections of papyrus scrolls and codex books coexisted in late Roman libraries, the destruction of the empire saw most of the great old library collections burned, plundered, or scattered. The codex book format proved hardier and more portable than the old scrolls; as a result, many scrolls failed to survive the fall of the Roman Empire, and a great deal of the literature that lived on did so between the covers of bound books.

By the fifth century AD, Rome had conquered much of continental Europe and most of the neighboring island of Britannia, dispersing copies of books that would eventually make their way into monastic libraries throughout the former provinces. For all of Rome's far-flung conquests, one island had always lain just beyond its reach: Ireland (or, as it was called, Hibernia). Throughout the era of Roman conquest, the Irish had lived in tribal communities not far removed from the preliterate cultures of Mesopotamia, Australia, or North America. Knowing nothing of reading, let alone Aristotle, people lived in small social groups with warring feudal kings and a shamanistic religious tradition, Druidism. Lacking any form of written language, they relied on oral traditions, mythology, and a web of symbols to preserve their understanding of the world. The old Druidic system was conceptually not far removed from the multilayered ancient folk systems of China, Greece, or Aboriginal Australia.

The Irish Celts were a famously bellicose people who warred with each other and occasionally raided neighboring England, where in the early fifth century a group of Irish pirates brought back a recalcitrant young slave named Saccath.

After working naked in the fields tending goats for several years, the slave managed to escape his captors and find his way to Gaul, where he spent the next twelve years at the monastery of Saint Germain, learning to read and studying the Gospels. Eventually, he decided to leave the monastery, vowing to return to Ireland to spread the word of God among his former captors. In the course of fulfilling that aspiration, the onetime slave—known to us now as Saint Patrick—would forever alter the course of Western European thought.

Even though Saint Patrick was not the first Christian missionary to reach Ireland, he was certainly the most successful. Within a mere thirty years, he managed to convert the entire Irish population to the new faith. Although Patrick's success undoubtedly had a great deal to do with his personal charisma, he also had technology on his side. Patrick introduced the Irish to the written word. He returned to Hibernia armed with a copy of the Latin Gospel. That lone volume would become the catalyst for Ireland's encounter with literacy. Patrick initiated the long process of transplanting the literate culture of classical antiquity into the shamanistic oral culture of Ireland. The result was an extraordinarily successful cultural adaptation. The drift of classical literacy to the remote island culture resulted in a beautiful new art form with lasting cultural consequences: the illuminated manuscript.

Although the Irish were hardly the first tribal culture to come in contact with the literary arts, their experience differed in one important respect from previous historical encounters with literacy. Before the emergence of the bound book, literate cultures had always introduced the technology of writing to nonliterate people in a linear format: in stone, clay tablets, or papyrus scrolls. The Irish were the first people to learn to read from books. Ever since, Irish literacy has been inextricably bound (so to speak) with the form of the book.

The sudden introduction of this new random-access form of writing to a previously illiterate tribal society sparked a collision of cultures that would yield dazzling literary by-products. In short order, the Irish managed to embrace both writing and books, while imbuing them with their own unique cultural sensibilities. In the coming centuries, Ireland would become a kind of literary research and development laboratory, populated by talented scribes with a knack for innovation, working together in an environment largely insulated from the continental ravages of the Dark Ages. For the next five hundred years, the Irish scribes not only preserved the classical texts of the old Roman world, but they also recorded their own indigenous Celtic mythologies and folklore, infusing bound books with their cultural heritage of storytelling and symbolism. Unencumbered by the institutionalized hierarchies of Rome and knowing nothing of government bureaucracies, religious institutions, or schools, the tribal people of Ireland took to literacy with zeal. Newly literate Irish monks began to navigate the great corpus of classical and scriptural knowledge. At first, the early Irish manuscripts were artistically primitive affairs, scratched out on poor-quality vellum (made from cows' stomachs) and stuffed with cramped and inconsistent handwriting. But the scribes learned quickly and soon showed a remarkable aptitude for the literary arts, mastering the finer points of majuscule and minuscule scripts, making notes in the marginalia, and decorating their texts with clever, sometimes provocative, illustrations. Within a century, they were producing manuscripts of astonishing beauty. These new kinds of books would become the gold standard of European illuminated manuscripts. Within two centuries, Irish scribes were producing breathtaking specimens like the *Book of Kells* and later the *Lindisfarne Gospels*. The illuminated manuscript became Dark Age Ireland's national art form.

For the scribes, copying manuscripts was an act of private devotion and contemplation rather than a rote task. Although the popular historical stereotype paints the monastic scribe as a dour clerical type, making laborious transcriptions—like the famous Xerox TV commercial caricature—the early Irish scribes were in truth anything but human copy machines:

> [The scribes] did not see themselves as drones. Rather, they engaged the text they were working on, tried to comprehend it after their fashion, and, if possible, add to it, even improve on it. In this dazzling new culture, a book was not an isolated document on a dusty shelf; book truly spoke to book, and writer to scribe, and scribe to reader, from one generation to the next. These books were, as we would say in today's jargon, open, interfacing, and intertextual—glorious literary smorgasbords in which the scribe often tried to include a bit of everything, from every era, language, and style known to him.[3]

The scribes found ample opportunities for exploration in making sense of these imported coded texts. Their capacity for inventiveness and experimentation stemmed in no small part from their near-total isolation from the Roman church, which elsewhere on the continent was beginning to establish a governing structure with restrictive rules for its burgeoning monastic orders and churchmen. But in Dark Age Ireland, the scribes enjoyed a great deal of artistic leeway.

The Irish scribes not only preserved countless important texts but in the centuries that followed also propagated those texts back to a continental Europe that had descended into near barbarism. Most of the great Roman libraries failed to withstand the ravages of the Dark Ages. And although a few monasteries managed to preserve important collections of church texts and works by classical authors, no one else came close to approaching the artistic innovations of the Irish scribes. Without the Irish contribution, Europe might well have lacked the intellectual fortitude to withstand the expansion of Islam in the early Middle Ages. Eventually, the Irish style would work its way back to the continent, and conversely, continental scribes would influence the Irish to adopt a more conventional style. Irish monks began to adopt the uncial script style of the continent, while the Irish techniques of lavish illustration and marginalia seeped into the continental manuscripts. The cultural mutations of the Irish manuscripts became part of Europe's cultural DNA.

The early Christian movement that began in Ireland soon spread across the Irish Sea to Scotland and then to England (and eventually, as far as Iceland and Italy). In contrast to the great institutional hierarchy of the Roman church that would eventually displace it, the early Celtic church was mostly a grassroots operation, self-organized, gender harmonious (monks could marry), and geared primarily around individual contemplation rather than institutional dogmatism. Although the Roman church would eventually flex its political, economic, and military muscle to bring the wayward Irish scribes (and their British colleagues) back into the administrative fold, during the critical period following the collapse of Rome, Ireland harbored a period of remarkable literary innovation that would have lasting consequences for European culture. The legacy of the early Celtic church, with its anti-hierarchical, individualist emphasis, would reverberate for centuries.

In the archetype of the Irish scribe, the original literary intertwingler, can we recognize a distant ancestor of today's blogger? Working outside of any traditional institutional hierarchy, engaging whatever topics interest him, interpolating words and images using a newly introduced communications technology, the scribe and the blogger seem like kindred literary souls. Although we should be careful of stretching the parallel too far, the introduction of new technology in both cases seems to have given rise to a new ethos of self-directed writing, holding the promise of entirely new forms of expression.

The medieval manuscript not only opened up the possibility of random access to the contents of a document but also gave rise to new forms like marginalia, inline illustrations, and layered type styles that allowed readers to engage medieval manuscripts in a manner not far removed from what we now think of as hypertext. The scribes recognized the importance of visual cues, fusing words and images to create juxtapositions of meaning beyond the mere concatenation of words on a page. The Irish style of manuscript, with its lower minuscule-style typeface (as opposed to the uncial style prevalent on the European continent) became the artistic signature of the early Celtic church. Irish scribes brought the style to England, where it flourished in the great monastery at Lindisfarne and became the foundation for the early English scriptoria that produced such works as the *Lindisfarne Gospel*.

One signature of biblical manuscripts of this era are so-called canon tables, symbolic arches containing cross-references between sections of the Gospel. Designed to resemble a church, the canon tables were both metaphorical (a symbol of God's word housed in the church) and practical reference tools enabling the reader to move between related sections in the text, a kind of illuminated hypertext.

Although historians of the book celebrate the artistic achievements of great manuscripts like the *Lindisfarne Gospels* and the *Book of Kells*, the vast majority of books produced in the early scriptoria were simpler affairs, intended for practical purposes. Scriptoria often produced durable copies of the Gospels to equip Christian missionaries on their journeys across the continent, where they used the texts to convert the pagans they encountered by wielding the power of God's "spell" ("Gospel"). The popular image of illuminated manuscripts suggests that they were the property of cloistered monks counting their proverbial angels on the heads of pins, but many early manuscripts went out into the world. Their physical heft and rich symbolism proved to be powerful physical totems against the disembodied oral traditions that held sway over the illiterate souls the missionaries intended to save. Often, when missionaries first encountered a pagan tribe, they would approach their prospective converts bearing a cross in one hand and a book in the other. To the illiterate tribes, the book carried far more power as a symbolic object than as a container of words, representing what Christopher de Hamel calls "tangible proof of their message" and a potent counterargument to ethereal oral traditions.[4] For the next thousand years, the monastic art of book making would play a central role in the spread of Christianity.

Although the books were primarily containers of written words, symbolism was integral to the manuscript form; it was through the power of images that the missionaries could captivate their illiterate audiences and then spellbind them with the ability to decipher the strange codes accompanying the images on the page. The Venerable Bede describes the importance of skillful imagery in the early spread of the Gospel, noting that the effect of pictures was "to the intent that all . . . even if ignorant of letters, might be able to contemplate . . . the ever-gracious countenance of Christ and his saints."[5]

As the craft of the manuscript progressed, the art of illumination became integral to the form. As a priest visiting Ireland in 1185 put it, "Look more keenly at [the illuminated manuscript] and you will penetrate to the very shrine of art. You will make out intricacies, so delicate and subtle, so exact and compact, so full of knots and links, with colours so fresh and vivid. . . . The more often I see the book, the more carefully I study it, the more I am lost in ever-fresh amazement, and I see more wonders in the book."[6] These books, full as they were of "knots and links," represented a weaving of written meaning and imagery that

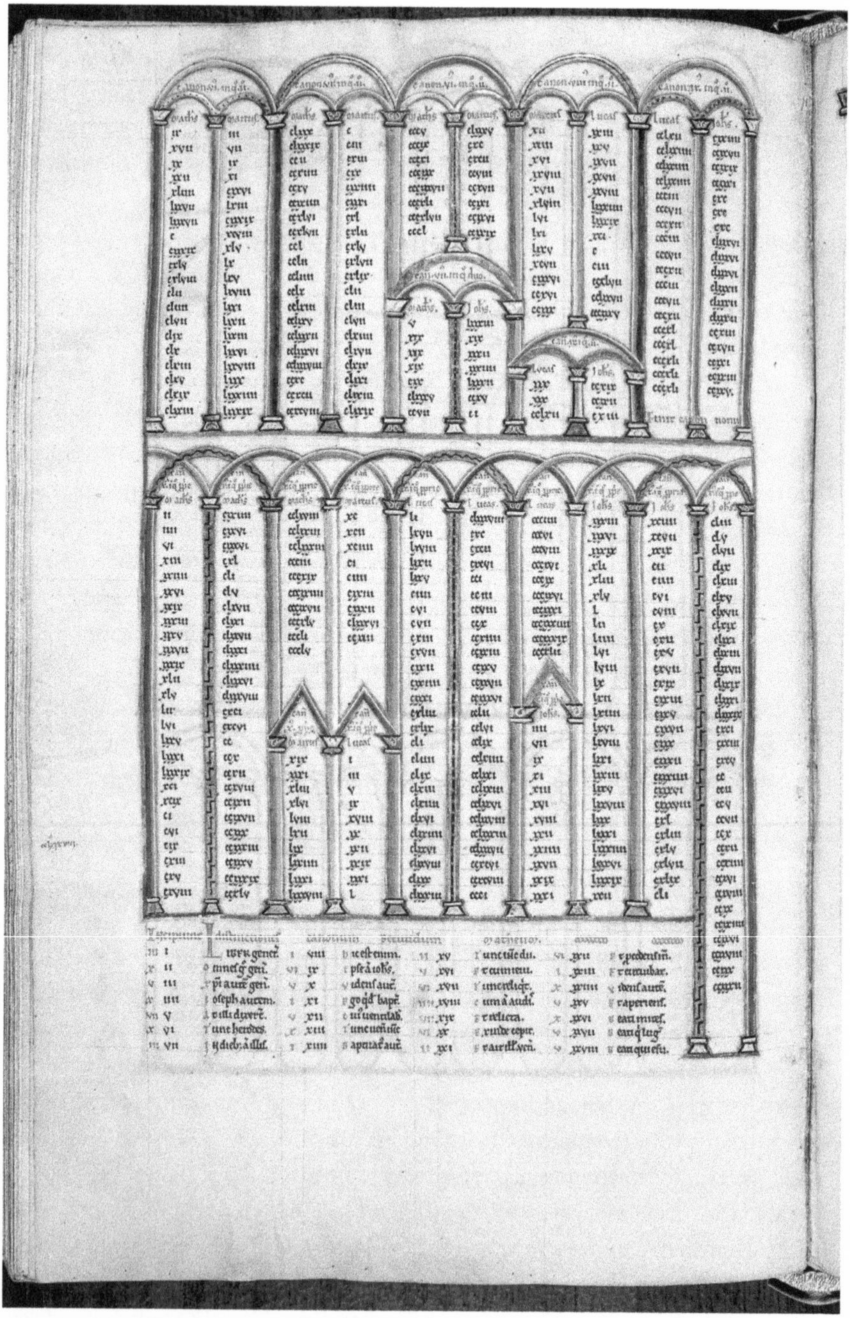

FIGURE 13. Page from a canon table in a Gospel from Helmarshausen, ca. 1120–40. J. Paul Getty Museum.

would flourish in the age before Gutenberg. Before automated printing ushered in an era of relentlessly textual books, the information architecture of the Dark Ages revolved around the union of word and image.

In the ancient archetype of the monastic scribe, perhaps we can recognize a distant ancestor of today's multimedia artisans. Like the self-directed artists of the scriptorium, today's generation of online writers and artists taking an unexplored new medium, the web, into their own hands, fusing words and images, interpolating old texts, and creating whole new forms of expression that, like the illuminated manuscripts of a millennium past, are brimming with vivid colors, knots, and links.

Houses of Mumblers

At about the same time that Saint Patrick was escaping his masters in Hibernia, a Roman nobleman named Cassiodorus was engineering an escape of his own. During the siege of Rome, he had witnessed the destruction of the last great Roman libraries, the Palatine and Ulpian. Although he stayed in Rome during the early years of rebuilding, he felt uneasy about the gathering political might of the church, ultimately deciding to forsake the ravaged city in search of a contemplative life. He abandoned the city and traveled to southern Italy, where his family maintained a country estate at Calabria. There, he decided to establish his own little monastery.

The little monastery would eventually tower over the European intellectual landscape, playing a crucial role in preserving the Roman literary heritage and pioneering techniques of book production and library cataloging that would reverberate for more than a thousand years. This was Europe's first great scriptorium. Although other European monasteries already had libraries of their own, Cassiodorus envisioned his scriptorium as more than just a literary archive. He saw it as an active center of scholarship and book production guided by an ethos of spiritual contemplation. He called it the Vivarium, "a place for living things."

Cassiodorus envisioned the scriptorium not as a musty archive but as an active center of learning, as a place where each monk "may fill his mind with the Scriptures while copying the sayings of the Lord." He also envisioned the scriptorium as a kind of missionary center, turning out copies of holy texts to be distributed across the growing expanse of Christendom. "What [the monk] writes in his cell will be scattered far and wide over distant Provinces,"[7] he wrote. The Vivarium soon became the most prolific center of book production the world had ever seen, churning out copies of works that eventually became canonical texts throughout Europe. Employing a small cadre of literate monks, Cassiodorus initiated a program of copying and translation that would become a template for a scholarly enterprise that flourished in European monasteries well into the Middle Ages.

In contrast to the inventive and sometimes freewheeling Irish scribes, Cassiodorus was a strict literary constructionist, holding his monks to a high standard of exactitude and consistency as they turned out uniform copies of the Gospels and other scriptural works as well as classical Greek and Roman writings. In their quest for efficiency, the monks pioneered new techniques of book production, introducing efficient new binding techniques and engineering a system for mechanical lighting that enabled them to work on their copying projects well into the night.

Cassiodorus did more than institute an efficient literary assembly line; he also postulated a philosophical framework that attempted to impose a higher order on the whole enterprise. Having fled the authoritative reach of the church and its hierarchical doctrines, Cassiodorus felt at liberty to propose his own system for classifying the growing body of works now in his charge. In his landmark work, *Institutiones divinarum et saecularium litterarum* (Foundations of divine and secular literature), he laid out a sweeping system of thought that proposed a unified organizational scheme for both scriptural and secular texts. Here, he proposed a new hierarchy of knowledge, starting with the titular dichotomy between divine and secular works. Within each top-level category, he proposed a mirrored ordering scheme. Believing that the pagan wisdom of the ancients contained seeds of God's truth, he chose to arrange his collection of books in a bipartite arrangement, juxtaposing the sacred wisdom of the scriptures against the practical knowledge of the ancients. Within the divine realm, the Bible occupied the top spot in the hierarchy, followed by the church fathers, then later and lesser commentaries. On the secular side, Homer occupied the lead position in the literary pecking order, followed by subsequent Greek and Roman poets, orators, dramatists, and historians. The system constituted, as Matthew Battles puts it, "a diptych of mirrored orderings of the divine and the worldly."[8]

Just as the ancient Romans had split their collections between Roman and Greek authors, Cassiodorus instituted a bifurcation between Christian and pagan writings that would reverberate for centuries in libraries and scriptoria throughout Western Europe. The system would become the foundation of medieval library catalogs for the next thousand years, serving as the de facto classifying scheme for most monastic libraries. Eventually, even the Vatican adopted Cassiodorus's scheme.

In addition to establishing a basic subject-level hierarchy, Cassiodorus introduced important innovations at the level of textual analysis. He established the practice of annotating texts in the marginalia and posited numerous lower-level classification schemes, like the one he devised for categorizing the psalms, which he divided into categories by the holy number of twelve:

1. The carnal life of the Lord
2. The nature of His deity
3. The multiplied peoples who tried to destroy Him
4. That the Jews should cease their evil ways
5. Christ crying out to the Father in the passion and being resurrected from the dead
6. Penitential psalms
7. The prayers of Christ, chiefly in His human nature
8. Parables, tropes, and allegories, telling the story of the life of Christ
9. Psalms beginning with the exclamation Alleluia
10. "Gradual" psalms, fifteen in sequence
11. The praises of the Trinity
12. Seven psalms of exultation at the very end[9]

At its peak, Cassiodorus's library numbered perhaps seven hundred books, a far cry from the seven hundred thousand scrolls at Alexandria. But what they lacked in volume they made up for in attention to craft. Given the limited resources of monastic scriptoria and lacking the great aggregating power of the old empire, the focus of libraries shifted away from the brute accumulation of knowledge toward perfecting the craft of reproduction and textual exegesis. The scriptoria were dedicated primarily to the preservation, not the production, of knowledge.

The creation of a codex book required the cooperation of several skilled workers: a skinner, a parchment maker, a beekeeper (for wax tablets), perhaps a painter, a book binder, and of course the ink-stained scribe. Often, monasteries would employ illiterate scribes to make copies of texts; they were considered more accurate copyists because they lacked the ability to inject their own ideas into the text.

The monks' relationship to books was deeply shaped by the oral culture that surrounded them. A page was not a flat slab of data but a living template for the spoken word and the refracted image. Understanding the medieval mind requires an imaginative leap. "For the monastic reader," writes the critic Ivan Illich, "reading is a much less phantasmagoric and much more carnal activity: the reader understands the lines by moving to their beat, remembers them by recapturing their rhythm, and thinks of them in terms of putting them into his mouth and chewing."[10] Indeed, monasteries were sometimes known as houses of mumblers. As Walter Ong puts it, "Manuscript cultures remained largely oral-aural

· Corncius.

e le bataulles manrenur.
c om les nes furct establia.
z a grans ustoue eles nauic

p oz fes par fon deununand.
e c an painfer an corment.
e ce uoy rcdum apzes.

FIGURE 14. Drawing of a monastery book cabinet, from a fourteenth-century edition of the *Roman de Troie* by Benoît de Sainte-Maure. Paris, Bibliothèque Nationale, ms. fr. 782, fol. 2v.

even in retrieval of material preserved in texts. . . . Readers commonly vocalized, read slowly aloud or *sotto voce*, even when reading alone, and this also helped fix matter in the memory."[11]

As the old hierarchies of the Roman Empire disintegrated into a period of chaos, new knowledge networks emerged at the hands of self-directed agents like Cassiodorus and the Irish scribes. Rejecting traditional political hierarchies (as personified in the Roman church), the monastic scribes prized individual intellect as the most reliable vessel of God's truth, pursuing new pathways and forging new ways of knowing from which new ontological hierarchies would eventually emerge.

Flowers from the Walled Garden

Throughout the Dark Ages, the monastic scribes protected countless classical texts and scriptural commentaries, preserving the flame of literacy while most of the continent reverted to the old feudal system. During this period, monastic libraries maintained their own small idiosyncratic collections of books. The monk responsible for tending the books was known as the *armarius*. In many monasteries, this monk held a position of high esteem, second only to the abbot. The typical monastic library owned at least one Bible, and most maintained editions of works by major theologians like Augustine and perhaps Jerome. They also collected a flotsam of lesser commentaries, classical poems, lives of the saints, local laws, and assorted popular lore. Books were difficult to come by, and so most of these collections remained small, typically numbering only a handful of volumes, rarely more than a hundred. Given the high costs associated with producing illuminated manuscripts, monasteries valued their books as prized assets. Monks kept detailed records of their holdings with written lists that looked more like an inventory of land holdings than a library catalog. Some of these inventories seem to list books in order of perceived value to the monastery, while others list them in chronological order, and others used organization schemes that remain entirely inscrutable. For example, the library in Würzburg, Germany maintained a list of thirty-four books, arranged in a seemingly idiosyncratic order starting with *The Acts of the Apostles*, followed by Gregory the Great's *Cura pastoralis* and *Dialogues*, a commentary by Jerome, Bede's ecclesiastical history, and theological works by Augustine and Ambrose. The list appears to follow a determined logic, but the structure of the catalog remains anyone's guess.

The forty-eighth chapter of the rule of Saint Benedict prescribes that each monk should read exactly one book per year, decreeing that "at the forty-day period [of Lent] they each receive a single book from the library, and each shall read the entire book from front to back; these books shall be given back at the beginning of the [next] Lenten season."[12] Monasteries guarded their collections closely; books were far too valuable for individual monks to own (with the notable exception of the Dominicans, who allowed their friars to keep small libraries).

Although the monastic scribes did their best to preserve the heritage of written thought inside the monasteries, outside the cloister most people had reverted to the old ways of oral folk traditions and symbolism. For most of the thousand years before Gutenberg, reading was an almost entirely lost art. In the eighth century, however, Europe witnessed a brief renewal of broader attempt at fomenting widespread literacy when King Charlemagne championed the cause of education throughout the so-called Carolingian renaissance. In 785, he issued his famous *Epistola de litteris colendis*, demanding that monks throughout the kingdom learn to read and write Latin and ensuring that each monastery have its own school.[13] He further attempted to make a public education available to all children under his rule, leading to a resurgence in public literacy. Although the movement would ultimately falter after Charlemagne's demise, it nonetheless triggered a renewed interest in books that would reverberate in the monastic scriptoria for centuries to come.

Like Ashurbanipal and Ptolemy before him, Charlemagne also recognized the political potency of a great imperial library. Unlike any other European king of

FIGURE 15.   Portrait of a scriptorium monk at work, from William Blades, *Pentateuch of Printing with a Chapter on Judges* (1891).

the era, he also seems to have harbored a deep personal passion for reading. Throughout his reign, he devoted enormous resources to building a library fit for an emperor. When Charlemagne needed to find a suitably learned man to carry out his great library, he looked across the channel to England, whose scribes had by 782 established the English monasteries as Europe's literary powerhouse. He invited a renowned scholar from York named Alcuin to his court, charging him with building the greatest library in Christendom. After arriving at Charlemagne's court, Alcuin soon sent word back to his old school in York: "I say this that you may agree to send some of our boys to get everything we need from there and bring the flowers of Britain back to France that as well as the walled garden[14] in York there may be off-shoots of paradise bearing fruit in Tours."[15]

Alcuin collected works by Statius, Lucian, Terence, Juvenal, Tibullus, Horace, Claudian, Martial, Servius, and Cicero, among many others. The English monk set about establishing not only a great imperial library but also a central repository for distributing template manuscripts to be copied and disseminated widely across the growing empire. To that end, Alcuin introduced a new simplified style of script, now known as Carolingian minuscule, that enabled monks to copy works more quickly to speed up the distribution of texts across the far-flung empire. The letters you are now reading on this page are the direct typographic descendants of Carolingian minuscule.

As the Carolingian libraries grew, monastic librarians began recognizing a need to keep better track of their assets. A few monks began listing their books chronologically. Eventually, some of them began creating higher-level categorical schemes, usually derived from Cassiodorus's system. At the Abbey of Saint Riquier, monks separated the books into discrete collections: one section for the Bible and theological works, one section for schoolbooks and grammars, and a third section for liturgical texts. "As humanism sought to recapture the sublimities of ancient Latin, on the one hand, and to authenticate Scriptures—and the power of the church—on the other," writes Matthew Battles, "its libraries recapitulated the symmetries of classical antiquity."[16]

Within each category, the books would be arranged in a rough chronology (alphabetical order would not emerge again until the twelfth century). As monks copied each other's texts, they became increasingly interested in the questions of librarianship: Which texts should they acquire, and how should they be organized? For guidance in developing their collections, monastic librarians would consult popular bibliographical works like Jerome's *De viris illustribus* and Cassiodorus's *Institutiones*—seminal guides to book collecting that would provide monastic librarians with a template for growing their collections. As these standardized bibliographies began to circulate among medieval libraries, they provided the foundation for an emerging scholarly network. Libraries began coordinating their collections, arranging them with similar structures and learning to rely on each other for access to hard-to-find works.

In 1170, the first known *scrutinium* (catalog) emerged—a reckoning of the monastery's collection of books. "With this invention," write Ivan Illich and Barry Sanders, "the book became dislocated from the sacristy."[17] In 1271, the nascent Sorbonne library created a catalog of its founding collection of roughly three hundred volumes, donated by the theologian Gérard d'Abbeville, which included a number of works from the library of Richard de Fournival, another noted book collector of the time and author of a catalog of books called *Biblionomia*. The library organized its collection into three sections: Philosophy, Medicine and Law, and Theology. Within the Philosophy section, it further subdivided its holdings as follows:

1. Grammar
2. Dialectic

3. Rhetoric
4. Geometry and Arithmetic
5. Music and Astronomy
6. Physics and Metaphysics
7. Metaphysics and Morals
8. Mélanges of Philosophy
9. Poetry

The 1495 catalog of the Carthusian cloister Salvatorberg in Erfurt stands as the largest known catalog of the late Middle Ages. In contrast to the Sorbonne catalog's subject-oriented scheme, it contains a list of authors arranged chronologically (rather than alphabetically), with each entry accompanied by a brief biographical sketch of the author, along with cross-references to other works by the same authors. These catalogs, all printed as codex books, provide a glimpse into the emerging semantic structure of medieval libraries.

In Rome, Pope Sixtus IV initiated the great expansion of the Vatican Library, establishing the role of *scriptores* or, roughly, curators of the collection (a position that has persisted at the library ever since). The Vatican Library catalog of 1475 offers a revealing glimpse into the evolving organization of knowledge in the world's most powerful institution. The library adopted Cassiodorus's convention of splitting its books into sacred and secular collections. Up one side of the great library ran a series of tables with books chained to them; on the opposite side ran a series of secular works that roughly paralleled the arrangement of the sacred texts. The library likely looked something like this:

The Vatican catalog's semantic balancing act would fail to hold up over time, but it marked yet another manifestation of the institutional impulse toward imposing hierarchical order on a body of recorded information—as a means of imposing ontological order on the otherwise cacophonous outpouring of human intellectual labor.

TABLE 4

SACRED	SECULAR
The Bible	Aristotle
Fathers of the church	Astrologers and mathematicians
Doctors of the church	Poets
Saints	Orators

FIGURE 16. Interior of the Library of Sixtus IV from *On the Vatican Library of Sixtus IV* (Clark, 1899).

The Arabian Legacy

Although the European monasteries played a critical role in preserving the heritage of classical Greek and Roman literature, their collections looked paltry in comparison to the great libraries that were taking shape a world away in the lands of Islam. "Bagdad in its glory abounded with libraries," writes the library historian James Westfall Thompson.[18] After assuming power in 786, the caliph Harun al-Rashid dispatched literary raiding parties across nearby territories, instructing his soldiers to commandeer every book they could find. He also funneled enormous resources into the production of books, establishing large-scale production operations where a small army of scribes would create copies of books en masse to populate the city's burgeoning libraries, including his own enormous personal collection, which would grow over the centuries to come into the legendary Grand Library of Baghdad (Bayt al-Hikmah).

Over the course of nearly five hundred years, the Grand Library evolved into more than just an archive; it also served as a powerful foundry of translation and knowledge production, with a flourishing international scholarly community. Arabic scholars worked alongside Syriac Christians, Zoroastrians, Persians, and Jewish scholars to translate texts and synthesize learnings from across these diverse traditions. Texts flowed to Baghdad from as far away as India, where important astronomical and mathematical texts by scholars like Brahmagupta, whose landmark work the *Siddhanta* introduced the concept of zero. The Persian writer Muhammad ibn Musa-al-Khwarizmi introduced foundational mathematical concepts like algebra, algorithms, and quadratic equations. And Ptolemy's landmark astronomy text the *Almagest* found new life in translated Arabic versions emanating from Baghdad that would ensure the preservation of Greek astronomical observations centuries to come.[19]

Before Baghdad fell to the Mongols in 1258, there were no fewer than thirty-six libraries in the great city, supplied by more than one hundred book dealers tending stalls in the great bazaars. The last vizier, Ibn al-Alkami, personally owned ten thousand books. The historian Omar al-Waqidi is said to have possessed 120 camel loads of books, while a scholar named al-Baiquai needed thirty-six hampers and two trunks to transport his collection. Legend has it the library at Tripoli held three million books, including fifty thousand Korans (although historians believe those numbers were probably exaggerated). In Cairo, the caliph Al-Aziz established a school library with 100,000 volumes (some say 600,000) bound books, including 2,400 Korans illuminated in gold and silver. The collection also included books on law, grammar, rhetoric, history, biography, astronomy, and chemistry. The library stored its collections in large presses behind a series of locked doors, each with a list of the books contained inside nailed into the front. In 1004 AD, the caliph Al-Hakim opened a so-called House of Wisdom in Cairo, said to contain 1.6 million books.[20] In contrast to the closely held literary treasures of the European monasteries, the caliph threw his library doors open to the public.

> The books were brought from the libraries . . . and the public was admitted. Whoever wanted was at liberty to copy any book he wished to copy, or whoever required to read a certain book found in the library could do so. Scholars studied the Koran, astronomy, grammar, lexicography and medicine . . . books in all sciences and literatures of exquisite calligraphy such as no other king had ever been able to bring together. Al-Hakim permitted admittance to everyone, without distinction of rank, who wished to read or consult any of the books.[21]

Whereas Europe languished in the Dark Ages, Islamic libraries flourished, preserving a huge swath of the heritage of classical thought. Greek, Hindu, and Persian books found their way by the thousands into the great libraries of Islam. The rise of Islamic libraries seems at first like an improbable outcome. Muḥammad did not read or write, and the indigenous Arab tribes had no written heritage. Traditionally, they were storytellers, preserving their legends, poetry, and genealogies through oral transmission. Thanks to a confluence of historical circumstances, these nomadic storytelling people became the curators of the ancient world's philosophical legacy.

Ancient books followed a circuitous path to the Arab world. When the emperor Justinian closed the great School of Athens, seven of its most prominent teachers went into exile, eventually finding refuge in Persia. They brought their books with them. Before long, Persia became a repository for Greek philosophy, poetry, and science. Hundreds of translators (mostly hellenized Syrians) gathered to translate the great works of classical antiquity from Greek to Persian. When the Arabs later conquered Persia, they assumed control of this great body of texts, which they proceeded to translate into Arabic. After centuries of isolation and illiteracy, the Arabs (like the Irish) experienced an abrupt, dramatic encounter with the literary heritage of the ancients. And like the Irish, that encounter would reshape their culture. For more than five hundred years, the Arabs went on to foster a period of scholarship that would fuel their rise to regional dominance and technological progress that made early medieval Europe seem, by comparison, like a cultural backwoods. As Edward Gibbon put it, "The age of Arabian learning continued about five hundred years . . . and was coeval with the darkest and most slothful period of European annals."[22]

In Baghdad, Cairo, and Córdoba, learned Jews played an important role in the expansion of Muslim philosophy, producing generations of diligent scholars who contributed to the expansion of learning in the Arabic world. The works of Hippocrates, Galen, and Ptolemy were preserved entirely from Hebrew sources. The Muslims generally treated their Jewish population well, following the Prophet's instruction to act indulgently toward their Jewish brothers.

With the Muslim conquest of the Iberian Peninsula, these translations found a toehold in Andalusia, whose rulers encouraged scholarly emissaries to travel widely in search of texts to bolster the young city's intellectual fortifications. The journey of the *rihla* became a rite of passage for many young Andalusian men. Such a journey was, as the independent historian Violet Moller puts it, "primarily a search for religious enlightenment, but, in reality, it often involved acquiring secular, scientific knowledge as well."[23] Thanks to these itinerant scholars, knowledge began to flow to Córdoba from the great centers of Islamic learning like Cairo, Baghdad, and Timbuktu—slowly amassing thousands of ancient texts in the Andalusian capital. The caliph Muḥammad I created an enormous royal library—by some accounts the largest in the world at that time—while wealthy merchants shored up their own private libraries in the great book markets that took shape on the streets of Córdoba.[24] Eventually, the library of Córdoba swelled to four hundred thousand volumes. This fertile intellectual climate created the conditions for the emergence of important scholars like Maimonides and ibn Rushd (also known as Averroes), widely regarded as the progenitor of the secular tradition in Western European thought.[25]

In Toledo, another important center of scholarship took shape as Jewish, Arab, Mozarabic, and Christian scholars worked side by side throughout the tenth and eleventh centuries, studying and copying texts in Arabic, Latin, and Hebrew. Works by seminal Greek thinkers like Galen, Euclid, Ptolemy, and Aristotle circulated and intermingled with texts by important Islamic scholars like al-Farabia and al-Khrawizmi. As Moller puts it, "The city was a bridge between Graeco-Arabic culture and Latin Europe, a place where scientific knowledge was not only held in safety, but translated and transmitted to the scholars of the future."[26] These energetic Islamic scholars preserved the legacy of classical thought across these Spanish centers of learning, until 1236 when the Catholic king Ferdinand II ordered the great library of Córdoba destroyed. After the Moors were expelled once and for all from Spain in 1492, the emperor Charles V ordered the burning of all books written in Arabic. In 1499, Cardinal Ximénez de Cisneros completed the purge of Islamic scholarship in Spain when he came to Granada and ordered an estimated 1.5 million books to be burned in raging bonfire in the city square. Fortunately for generations of scholars to come, a great many of the most important of these texts had long since been translated from Arabic to Latin, copied many times over, and safely dispatched to new repositories across the European world. But many more were lost. The Mongol invasions left many of the great Muslim libraries in embers. Those texts that survived long enough to resurface in Europe, however, would help seed a new respect for science and a revival of ancient wisdom that would ultimately pave the way for the great philosophical awakening of the European Renaissance.

A STEAM ENGINE OF THE MIND

How quiet the writing, how noisy the printing.

—Marina Tsvetaeva

In Victor Hugo's *Notre Dame de Paris*, a scholar sits in his study carrel inside the great cathedral, contemplating a strange new artifact that has just landed on his desk: a printed book. Surrounding the new book are piles of old illuminated manuscripts. The scholar walks to a nearby window and gazes out on the silhouette of the great cathedral. "*Ceci tuera cela*," he says. This will be the end of that.

Just as the advent of alphabetic writing caused massive social disruption across the ancient world, so the new technology of the printing press would transform the social and political worlds of medieval Europe. The Gutenberg revolution would destabilize the old feudal and religious hierarchies of the Middle Ages, enabling a growing class of literate Europeans to reach out to each other in new ways, forging new social bonds that would eventually culminate in the cultural transformations of the Renaissance.

Although traditional Western history has anointed Johannes Gutenberg with great man status as the inventor of movable type, the technology was not the invention of a lone European genius but rather the evolution and refinement of a number of precursor technologies, many of them emanating from Asia.

The Chinese invented the first known movable type system in 1040, using ceramic characters set in an iron frame. By the thirteenth century, a metal movable type system had appeared in Korea for printing ritual texts. The historian Donald F. Carter has speculated that these were not merely coincidental developments but that the Asian movable type technologies directly influenced the subsequent development of movable type in Germany.[1]

By the time Gutenberg introduced his famous press in 1458, the revolution that would bear his name was already well under way. Although the advent of mass printing would accelerate the transformation of Europe's information infrastructure, the seeds of change had already been planted inside medieval monastery walls. To understand the deeper trajectory of the Gutenberg revolution, we need to begin by stepping back a few centuries to the late Middle Ages, when the first glimpses of popular literature were starting to make their way out of the cloisters.

Why the Elephant Had No Knees

In early medieval England, one of the most popular books circulating among the middle and lower classes was a little volume known as the *Physiologus* or bestiary. Consisting of simple verse descriptions of animals, usually accompanied by colored illustrations, the bestiary tried to explain basic points of Christian doctrine in allegorical terms by presenting them in the guise of a book about animals. By the late twelfth century, it had become the most popular nonbiblical manuscript of its time—indeed, the first such work to reach a wide lay audience. By the early fourteenth century, almost every monastery and church parish owned a copy; schoolchildren learned it by rote in their classrooms. Ever since the first folk taxonomies emerged, human beings have been naming and categorizing animals. So perhaps it should come as no surprise that the bestiary should prove such a runaway best seller in pre-Gutenberg Europe.

The work had originated in the early Greek Christian community, persisting through the ages as successive generations of monks copied the manuscript and passed it around from monastery to monastery throughout early Christendom. Its origins stretch far back into classical antiquity. The textual material evolved from a seminal Greek work known as the *Physiologus* (meaning, "the naturalist"), whose bibliographical lineage stretches through a number of other books dating back to the reign of King Solomon. Parts of the book can be traced back to Pliny the Elder and Aristotle, some even further back to stories of Indian, Egyptian, and Jewish origin. In other words, the bestiary contained a healthy dose of folk wisdom from ancient oral tradition. By echoing the themes of popular folk taxonomies and portraying animals as symbols of other divine truths, the bestiary may have struck a deep unconscious chord with its illiterate audience who may have felt the stirrings of an ancient wisdom tradition at work.

The book was full of allegorical stories, using the figures of animals to tell a larger story about Christian theology to people who lived, for the most part, in an oral culture. We have seen how folk taxonomies give rise to synaesthetic

FIGURE 17. Adam naming the animals, from the *Aberdeen Bestiary*, Folio 5r.

knowledge systems that persist with remarkable fidelity in oral cultures. In the oral culture of the late Dark Ages and early medieval era, the bestiary tapped into this vein of folk wisdom while at the same time serving as a kind of boundary object between oral and literate cultures, introducing illiterate audiences to tales that were undoubtedly meant to be heard aloud yet were nonetheless obviously encased between the covers of a codex book.

The bestiary almost always begins with the image of Adam naming the animals, making a biblical claim for humanity's divine right to name—and, by extension, categorize—the beasts of the earth. "To name is to control," as Whit Andrews puts it, "and to establish primacy and the right of exploitation."[2] In this image, Adam not only names the beasts but also appears to assign a rudimentary taxonomy, allotting each beast to a particular box in his scheme. Each animal's name includes a divine etymology, with families of animals arranged by both family and relative proximity to humankind, with animals "not in man's charge" (the great cats and other wild quadrupeds), followed by animals "for use by men" (e.g., horses, cows, and sheep), and finally "beasts" (e.g., dogs, cats, and boar). This image also echoes the folk classifications of preliterate societies in which affiliate classes provide a category for grouping animals by their relationship to human beings as well as by their relationships to each other.

The author (sometimes also known as Physiologus) attempts to draw instructive moral lessons from the behaviors and qualities attributed to various real and mythical animals. Each chapter conforms to a simple pattern: first, an animal is described in physiological detail; then its properties are shown to illustrate a particular theological point, often buttressed with biblical passages. Elephants were said to have no knees; when they slept, they had to lean against a tree, and the tree in turn symbolized "tree of life," or the pillar of faith. Lions were said to give birth to stillborn cubs which came to life only on the third day after their birth, symbolizing the Resurrection. And so on. The *Physiologus* made no pretense at scientific accuracy; these were allegorical tales, meant to illustrate symbolic rather than literal truths.

The *Physiologus* traveled through successive generations of copyists from ancient Greece to libraries as far flung as Ethiopia, Armenia, Syria, and Rome. It was in its last translation to Latin that the text took on a relatively stable form that would form the template for a thousand-year run in successive copyists' hands across the monasteries of Europe, eventually finding its way into Italian, French, German, Spanish, and English. During this period of northward migration, the work passed through the hands of generations of monastic scribes, evolving along the way to incorporate local nuggets of animal lore and superstition. With each successive copying, both text and image would change slightly as scribes made alterations from the original. One copy served as the basis for the next, and hundreds of small alterations eventually accumulated into signifi-cant changes. By the eleventh century, the work contained descriptions of almost double the number of beasts found in the original *Physiologus*, evolving over time through a long process of mutation. Eventually, these iterations turned the work into a whole new book, no longer called the *Physiologus* but instead the *Besti-arum* (Animal book). By the time the genre reached its English heyday, it began to expand toward a protoscientific form, including entries that found their way into the work with no apparent religious significance. Isidore of Seville's famous *Etymologies*, for example, contained descriptive materials similar to those found in the *Physiologus* but without the Christian allegories. In these larger works, doctrine and religious interpretation had to share the stage with a more secular and straightforward view of the natural world. Thus, the bestiary began paving the way for the more formal classification systems that would come later at the hands of naturalists like Carl Linnaeus and Comte de Buffon (see chapter 9).

Popular demand for copies of the bestiary fueled the development of inven-tive new reproduction techniques in the monastic scriptoria, methods that in turn allowed the work to reach even wider audiences. English bestiaries of the thirteenth century bore remarkable similarities to one another. The text and il-lustrations mirrored each other so closely, in fact, that book historians have surmised that they must have been produced from common master copies. The practice of using such "model books" for the production of popular manuscripts flourished in European scriptoria of this period. In the well-known K manu-script, illustrations reveal small puncture marks tracing the periphery of each figure, along with one hole in the middle of the page—some pricked so force-fully that they penetrate through to the following pages. Samuel Ives and Hellmut Lehmann-Haupt have tried to trace some of these techniques through a careful examination of the physical evidence: "The central hole in the miniatures prob-ably served to keep the leaves in exactly the same position during the punctur-ing process. After they were pricked, the loose sheets were laid onto the pages of the new copies of the Bestiary that were being produced at the workshop. Pow-der was dusted through them, the dots [were] joined, forming a fine outline drawing perhaps in light pen or thin brush stroke. This was then painted in by illuminator, using the originals as color guides."[3]

This process of replication, known as the punch transfer technique, marked an important breakthrough in the technology of book production. Now scriptoria could produce copies of the same work with much greater speed and accuracy. Although the technique was a far cry from the automated processes of Gutenberg, it marked a major leap forward in medieval information technology. Most of the surviving bestiaries from this period have a hurried, inelegant quality to them—further evidence of mass production. Although a few finely wrought specimens survive, the majority bear the mark of hurried production, with spare illustrations and sometimes unbeautiful handwriting. Most of the surviving bestiary manuscripts of this period feature only a single illustration per passage, while earlier versions contained many more.

In the age just before Gutenberg, the technologies that facilitated the distribution of bestiaries placed the book in a central role in both technological and cultural change. Bestiaries became an important forerunner of later popular books, in both form and content. The images they contained became a cornerstone of popular Christian iconography for centuries to come, cropping up regularly in the church architecture of the day as sculptural adornments in even the simplest of parishes, where illiterate parishioners would listen to sermons and Gospels and then find those lessons reinforced in the visible symbols of the church architecture. Bestiaries also figured prominently in the medieval school curriculum, taking a place as one of the standard instructional texts along with Cato, Aesop, and Prosperus. This entrenchment in the popular nonliterate psyche explains the hold these images would later have on successive English generations. The images so imprinted themselves on the popular consciousness that they eventually appeared in tapestries, paintings, mosaics, and even fabric cloth. English authors from Edmund Spenser to William Shakespeare to C. S. Lewis regularly drew on animal imagery that can be traced directly back to the bestiary.

Long after the *Physiologus* faded from the popular fancy, the images and themes it contained persisted, insinuating themselves into the symbolism of later ages. The bestiary's legacy extends beyond our lingering fascination with imaginary animals, however. Its success helped fueled a popular demand for books that would plant the seeds of the Gutenberg revolution.

Toward Literacy

As Europe entered the Middle Ages, the production techniques perfected in the monasteries fueled a rising popular interest in books. Once again, information technology was driving cultural change. Although literacy was slowly beginning to spread among the monastic and aristocratic classes during the late medieval period, the vast majority of the population remained illiterate. Folk wisdom still predominated, and oral traditions continued to flourish. But the success of the bestiary suggested a new public role for books and writing as more than just instruments of monastic scholarship but as vessels of popular wisdom.

Across Europe, literacy was reemerging along the contours of a familiar historical pattern. Just as writing in the ancient Near East had emerged at first in the mode of "craft literacy," supporting commercial transactions, so the reemergence of writing in Europe may have had at least as much to do with Mammon as with God. As the historian Brian Stock writes, medieval Europe's reencounter with literacy "would seem to have repeated with minor variations a process that unfolded in the eastern Mediterranean centuries before." Just as in Mesopotamia, "the rebirth of medieval literacy coincided with the remonetization of markets and exchange."[4]

Charlemagne had contributed to the secularization of writing by forbidding priests from conducting business transactions, such as drawing up contracts between illiterate parishioners. With priests barred from creating documents, a new class of secular scribes began to emerge. Coupled with a revival of Roman law in Italy that began to spread across the continent, the newly privatized class of scribblers found a growing market in writing up quotidian documents, including various bonds, pacts, and legal agreements. The privatization of secular writing also yielded an economic incentive for the scribes to market their services. Whereas in the Dark Ages, Europeans had largely done business on a word and a handshake, by the fourteenth and fifteenth centuries they increasingly could avail themselves of willing scribes who, for a fee, would write down the details of their business arrangements.

Throughout the Dark Ages, most people conducted transactions requiring close personal bonds of trust, like the highly ritualized pledges of fealty between vassals and lords (often sealed with a kiss on the lips). Now these old forms of intimate, person-to-person social contracts started giving way to more detached forms of written agreement. Although the vast majority of the population remained illiterate, the technology of writing exerted a growing influence over people's lives. "When written models for conducting human affairs make their appearance, a new sort of relationship is set up between the guidelines and realities of behavior," writes Stock. Such documents introduce consensual rule sets that constrain social actions. "Moreover, one need not be literate oneself in order to be affected by such rules. A written code can be set up and interpreted on behalf of unlettered members of society, the text acting as a medium for social integration or alienation."[5] In other words, literate civilization is not necessarily predicated on mass literacy.

As written laws and contracts began to proliferate throughout Western Europe in the eleventh and twelfth centuries, the function of writing started to change. Documents were no longer just vessels of information—like the word of God, the poems of Virgil, or tales of fanciful beasts. A new class of documents emerged that functioned less as narrative texts and more as facts unto themselves. "Men began to think of facts not as recorded by texts but as embodied in texts, a transition of major importance in the rise of systems of information retrieval and classification."[6] In a society where most people still could not read, the document served less as a text and more as an externalized "social fact" (a term coined by Émile Durkheim): its mere existence often mattered more than the particular words it contained.[7] Even today, in an age when most people can read, we still treat many documents this way. For example, how many of us ever read the words in a warranty, a stock prospectus, or a software license agreement? Such legalistic documents typically serve the same function as medieval contracts insofar as they matter more as symbolic seals of trust than as conveyors of textual meaning.

The rise of social documents in the early Middle Ages fueled an expansion of commerce and trade, laying the groundwork for a series of larger social, political, and economic transformations to come. Just as symbolic objects like beads and bone knives had transformed and expanded existing social structures facilitating a "release from proximity," the new technology of secular documents allowed people to forge bonds of trust among people otherwise lacking any social connection. Like tribal beads and jewelry, medieval documents functioned primarily as totemic objects, used to cement social relationships that could extend beyond any one individual's immediate trusted social network. Now two relative strangers could forge a reliable bond in the form of a document, even though in all likelihood neither party could read it. Merchants from distant towns could make pacts with each other; treaties, laws, and institutional charters began to emerge. Whereas writing had once been the exclusive province of holy men, now it passed into the hands of the laity. Merchants had contracts drawn up, government officials documented their laws, and increasingly people learned to govern their own affairs under the aegis of the written word. Even though most people still could not read, documents provided the conduit for medieval Europe to move toward larger and more complex social and political forms.

The technology of writing had been locked for centuries in the monasteries. To medieval Europeans, it was an instrument of divine power, a secret accessible only to holy men. As secular literacy started to spread beyond the monastery walls, written words still carried a residual talismanic authority that lent documents an extraordinary symbolic power. The once-mysterious art became part of the fabric of everyday social life.

Late-medieval writing was, then, a form of stigmergy (see chapter 1). Just as ants and termites transform the external environment to introduce coded constraints on each other's behaviors, so medieval documents served a similar purpose as totemic objects designed to regulate behaviors; their function had less to do with semantic meaning than with social coercion. "Relationships between the individual and the family, the group, or the wider community are all influenced by the degree to which society acknowledges written principles of operation," writes Stock. A shared recognition of the power of documents can mediate and shape human relationships, whether or not every individual actually knows how to read. This kind of macro literacy "also affects the way people conceptualize such relations, and these patterns of thought inevitably feed back into the network of real interdependencies."[8]

As Europeans found themselves living in an increasingly document-centric world, they started to experience the first rumblings of social and political transformations that would reverberate for centuries. Whereas the changes that swept Europe during this period arose from a complex web of causes and conditions, the new technologies of writing played an important role, helping speed the demise of the old feudal system and fostering a period of turmoil that would reach an apex during the Gutenberg era. Ultimately, the impact of new technologies helped spur the growth of new forms of governmental and institutional hierarchies that would take full expression in the Renaissance. Eventually, the technology of secular documentation would even give rise to an entirely new form of government: the document nation, a state founded purely on the written word (whose greatest exemplar would eventually be the United States).[9] We are getting ahead of the story now, but suffice it to say that even before Gutenberg invented his press, a medieval information explosion was well under way, and it would prove literally revolutionary.

As secular writing emerged among the mercantile classes, literacy spread. Books were getting cheaper to produce, and more people could read them. The monastic scriptoria, which had once enjoyed a virtual monopoly on book production, were beginning to compete with an emerging institution of secular literacy: the university. During this period leading up to Gutenberg, books began to find an audience among the bourgeoisie, and Europe began to feel the first rumblings of an emerging secular literature.

The growth of universities, coupled with cheaper production materials, meant that monasteries lost their monopoly on the production of books. Now a new class of scholars, teachers, and students began to produce manuscripts of their own, often working with lay craftsmen who began to form a healthy and growing book trade, populated by booksellers, professional copyists, and "stationers."[10] The growing number of secular tradespeople involved in book production led to a flowering of more and more books. Universities began to accumulate books in growing libraries of their own. And as the universities turned out graduates, a new reading public emerged, fueling a market for books whose subjects extended beyond the time-honored genres of scriptural texts and works of classical antiquity. New types of popular texts, written in local vernacular, began to surface: travel journals, poems, romances, the lives of the saints, and the popular book of hours started to grace the homes of many medieval households.

During a period stretching roughly from the thirteenth to fifteenth century, a preliminary information explosion began to take shape, fueled by the introduction of a breakthrough information technology: paper. Although the old materials of parchment and vellum had persisted for centuries, the invention of pulp-based paper made it suddenly possible for scribes to produce cheaper editions of popular works, which a growing merchant class could afford to purchase for themselves. The cost of raw materials declined (although the cost of labor did not), and as a result, books started to proliferate among a wider audience.

By the end of the fourteenth century, an emerging literate public was consuming books in growing numbers. By the early fifteenth century, books were already being mass produced throughout Europe, through monastic scriptoria and other private sector workshops capable of turning out hundreds of books at a time. Fueling the spread of writing were innovations like woodblock printing and the punch transfer technique perfected in distributing books to the missionary corps. Demand for books was growing, and the growing corps of secular scribes were finding it difficult to keep up. Conditions were ripe for a technological breakthrough.

A Rumor in Mainz

In 1458, the king of France dispatched one of his spies to the German town of Mainz, where rumor had it that a troop of printers had created an ingenious new device, capable of reproducing manuscripts mechanically. No scribe needed.

The historical impact of Gutenberg's printing press requires no corroboration here. Suffice it to say that without the advent of movable type, we would live in an unimaginably different world. Today, we instinctively view the arrival of the printing press as an unqualified boon: the engine of democracy, modern scholarship, and individual expression. For all its positive by-products, however, the history of the printing press turns out to be far less rosy than we might imagine.

Gutenberg's machine arrived in Europe not as a benign force for personal enlightenment but as a profoundly disruptive technology that triggered a series of painful and often bloody conflicts. The arrival of the Gutenberg press may have ultimately heralded an unprecedented boom in reading and writing, but it also serves as a cautionary tale of the potentially devastating effects of a disruptive information technology.

Just ten years before the French king dispatched his spies, Gutenberg had invented his movable type system, signaling a staggering technological leap forward from the block printing technology already in use. Trained as a metalworker, he created an innovative new metal alloy that enabled him to create small metal letters capable of retaining ink on the surface long enough to make an impression on a sheet of paper. By pressing these letters together into a frame and coating them

FIGURE 18.   Press invented by Johannes Gutenberg (ca.1398–1468) in approximately 1436. © Bridgeman Images.

with ink, he could then reproduce entire pages of books with a single stroke. Using a modified cider press, he could then create multiple copies of a page in a fraction of the time that a quill-wielding scribe would have needed. After a few years of trial and error, Gutenberg perfected his technique and set about producing 180 copies of the book that would enshrine his name in history: the Gutenberg Bible.

Recognizing the potential value of his invention, Gutenberg had sworn his printers to the medieval equivalent of a nondisclosure agreement, making them promise to breathe not a word of it outside the print shop. It was a futile promise. Word of the invention soon found its way out of Mainz and into neighboring municipalities. By the time the rumor reached the king of France, word had already started to spread throughout continental Europe. By 1475, printing presses had popped up as far away as Paris, Lyons, and Seville.[11]

In the first decades of printing, the press found its most eager supporters among the clergy, who recognized the obvious labor-saving value proposition and rejoiced in the prospect of expanding the reach of God's word through the promising new technology. Scholars, clergy, and other literate people greeted the new technology effusively. "What an ascent toward God! What ecstatic devotion must we feel on reading the many books which printing has given us!"[12] Many of the first printers were themselves monks, eager to use the new device to spread the word of God far and wide among their fellow monasteries. Little did they realize that this useful contraption would soon play a role in their own institutions' ultimate downfall.

In the fifteenth century, the Roman church exerted a near-monopoly on the New Testament, carefully controlling access to the Bible and other holy texts. Even within the church, only a subset of elevated priests possessed the skill to read Latin and thus interpret the word of God. Churches and monasteries often kept their copies of the Bible chained to a desk or pulpit, or locked in a cabinet. These prized holy books were not just vessels of the word of God; they also carried a totemic power as symbols of church authority.

At first, the Roman church enthusiastically endorsed the new printing process, recognizing the potential economy of scale in distributing holy texts to far-flung churches and monasteries. Early printers would often set up shop at the behest of nearby cathedrals or monasteries, who subsidized their setup costs so that they could provide ready access to copies of the scriptures and other liturgical texts. The early printing business also flourished in part thanks to a bull market in printed indulgences,[13] which now became an increasingly profitable business thanks to the printing press's economy of scale.

The booming printing business soon proved too lucrative a business for the clergy to support by themselves; as the volume of printed documents grew, printing required an increasingly specialized workforce. Private presses were springing up all over Europe, especially in the growing university towns. In larger towns, there was an enormous untapped market of wealthy individuals eager to acquire texts previously accessible only to churches and governments able to employ armies of scribes. These wealthy patrons, men like Jean de Rohan (himself a man of letters), functioned much like venture capitalists today. They invested in the presses both for their own enrichment and out of a genuine interest in the propagation of knowledge. In large towns with an academic population, popular volumes included secular works by Aristotle, Aquinas, William of Ockham, and Boethius, as well as traditional religious texts like the works of Saint Augustine and other church fathers.[14]

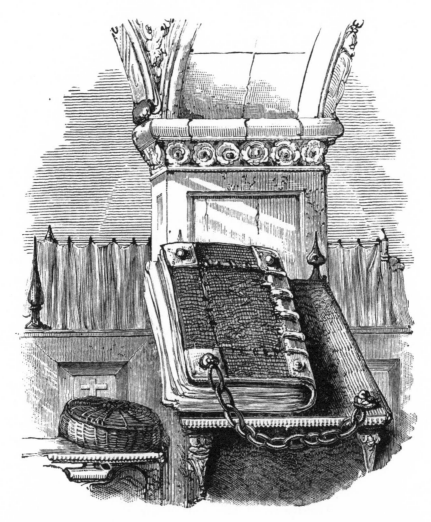

FIGURE 19.   Chained Bible in Cumnor Church, England. From William Andrews, *Curiosities of the Church (1891).*

Depending on local literary appetites, publishers would tailor their offerings to the tastes of particular market segments. Scholastic books were typically printed in black gothic letters. For the common lay reader, vernacular narratives were printed in bastarda gothic. And for an emerging class of readers that today would be called humanists, classical texts were printed in the new "littera antique" style, the ancestor of today's familiar roman typefaces (like the letters on this page).[15]

Soon, the emergence of printed books led to a standardization of typefaces that began to extinguish the idiosyncratic local styles that had flourished in the age of illuminated manuscripts. An alphabet soup of local typefaces gave way to the new roman style. Many scholars came to believe that roman type evoked the clean, economical style of the classical texts they so admired; they regarded gothic texts as overly ornamental, a throwback to old monkish ways. The adoption of roman typeface was more than just a stylistic decision; it also signaled a rejection of medieval scholasticism in favor of the secular wisdom of the ancients.

The standardization of typefaces would prove the catalytic event in the subsequent movement of information across cultures. Roman type became the equivalent of ASCII type today: a universally recognized standard that speeds the flow of information across a far-flung network of readers and writers. Within a century of Gutenberg's invention, roman script had taken hold across Europe (although printers in Germany continued to produce works using gothic typefaces, a preference that never completely subsided). Standard typefaces soon led to another form of standardization in the form of the printed book itself. New conventions took hold around the use of "meta"-information like title pages, colophons, and framed title pages. These conventions emerged over time and slowly coalesced into standard practices for European printers. The introduction of a convention for bibliographic metadata reverberates to this day; indeed, the title page of this book (and of almost every commercially published book) is the direct heritage of a five-hundred-year-old convention forged in the first few years after Gutenberg.

The rapid growth of the popular book market split the book trade into two broad segments: (1) large folio works intended for institutional collections like libraries, schools, and churches and (2) smaller quarto or octavo works for the public at large. This broad market segmentation persisted well into the nineteenth century, contributing to a growing schism between "canonical" and popular knowledge.

By the sixteenth century, mass-produced books were starting to reach a growing population of customers who themselves could barely read. This interim period between the emergence of printing and widespread popular literacy marks the heyday of the illustrated book. Books were increasingly within reach of a public eager to consume books that did not necessarily demand reading. Popular volumes often included simple line drawings accompanying the text, illustrating stories from the Bible, pictures of saints, or the terrifying visages of demons trying to wrest men's souls.[16] These sturdy xylographic illustrations were simple and clean, intended to serve as informational graphics rather than decorative illustrations.[17] "Book illustration answered a practical rather than an artistic need," write Lucien Febvre and Henri-Jean Martin, "to make graphic and visible what people of the time constantly heard evoked."[18] Like the widely distributed bestiaries, these popular books appealed to readers and nonreaders alike. The popular books of this era mark a brief period in which word and image still enjoyed equal prominence on the printed page.

The golden age of illustrated texts in the sixteenth century would not last. Growing popular literacy, coupled with inexorable logic of the marketplace, forced efficiencies in both the production and the consumption of books. "What the steam engine does with matter," writes John Newman, "the printing press is to do with the mind; it is to act mechanically, and the population is to be passively, almost unconsciously enlightened, by the mere multiplication and dissemination of volumes."[19] Whereas monastic scribes had produced books almost exclusively for religious purposes—to equip missionaries with books for their travels or to preserve the word of God for their fellow brethren—the printing press transformed book publishing into a for-profit industry. Fifteenth-century publishers, like their descendants today, would only publish books if they felt reasonably assured of turning a profit. The old craft production values of the monastic scriptoria could never survive in the new world of for-profit publishing.

At first, publishers turned out great numbers of Bibles and other devotional texts. In the thriving Jewish communities of Italy and Spain, Hebrew printers started publishing traditional Hebrew texts that were welcomed enthusiastically in their communities, where the new technology of printing afforded the possibility of reproducing the dense layered commentaries of Torah scholars.

Eventually, a public appetite for reading fueled the growth of an increasingly secular book trade: calendars, almanacs, and works by classical authors started to appear on bookstore shelves alongside popular religious texts like the book of hours or the lives of the saints. With a saturated market for traditional religious books, publishers started looking for new profit centers. "Under the mounting flood of new books written for an ever increasing public," write Febvre and Martin, "the heritage of the Middle Ages lost its hold."[20] By 1500, Gutenberg and his professional descendants had already published an estimated eight million books; by the end of the sixteenth century, the number stood close to two hundred million.[21]

Publishers were turning out a staggering number of volumes in local vernacular languages, starting with translations of classical authors like Virgil, Ovid, Caesar, Suetonius, and Josephus. Popular contemporary authors started to emerge in Italy, France, and Spain. Writers like Machiavelli, Erasmus, and Rabelais started to find a wide readership, while a new crop of scientific writers like Copernicus started to gain a following in scholarly circles. Books about geography, natural science, and medicine started to circulate. History books were enormously popular. "Gutenberg's invention produced what we today understand as scholarship," writes George Landow. "No longer primarily occupied by the task of preserving information in the form of fragile manuscripts that degraded with frequent use, scholars, working with books, developed new conceptions of scholarship, originality, and authorial property."[22] Whereas once the whole business of manuscript production had involved an intimate, largely anonymous collaboration among countless scribes, the press would introduce a new fixity, both to the text and to the concept of the author.

One popular travel book captured the reading public's fancy: Christopher Columbus's description of his voyage to America (published simultaneously in Barcelona, Paris, Rome, and Basel). By sixteenth-century standards, it was a runaway best seller. Yet lest we start to overestimate the impact of the printing press on the daily life of the average European, we should bear in mind that it still took the better part of a hundred years for news of Columbus's "discovery" of the New World to reach most people in Europe beyond a small circle of literate middle- and upper-class readers. For the most part, news still traveled by word of mouth. But that was starting to change.

This was the third great turning of the information explosion. In tribal societies, human cultures had been bound together by unitary knowledge systems in which everyone participated; in the age of city-states and empires, a literate scholarly class had emerged as caretakers of institutional knowledge, while the masses relied on oral traditions, imagery, and socially transmitted knowledge. Oral and literate cultures twirled in separate orbits. This third axial period of information technology would bring those orbits closer together, with occasionally cataclysmic results.

Just as primordial bacteria had walled themselves off from each other to form a new kind of self-directed organism, the new printed book signaled a consolidation of form. Oral traditions existed in a fluid medium, where authorship and provenance were fuzzily delineated. As the printing press took hold, those traditions would find their way into print and become more fixed in the process—forming nucleated texts with, as it were, thicker cell walls.

Handmade manuscripts had been fluid affairs too, produced by teams of craftspeople working in close collaboration to produce labor-intensive intellectual artifacts. Scholarship took place within tightly confined institutional constraints. Now the old bonds of collaboration and careful engagement had been sacrificed for the virtues of speed, lateralization, and efficiency, and soon scholars would find new ways of communicating with others across institutional barriers.

The new fixity of texts not only led to emerging forms of scholarship but also had both a galvanizing effect on hierarchical institutions—which drew great strength and power from the new invention—and a disruptive effect, as a newfound ease of communication allowed new kinds of public social networks to emerge. Stock has written of how the spread of documents and literacy led to the formation of "textual communities" that coalesced around particular documents, especially religious and political movements that challenged the established organizational hierarchies. Without the printing press, Martin Luther could never have published his ninety-five theses or Thomas Jefferson circulated the Declaration of Independence.

The text became a shared object, a replicable unit of information that could move freely through the social body of the literate populace. Textual communities flourished in an era of newly fixed documents in which meaning was no longer interpreted through a scholastic elite possessed of secret knowledge but freely disseminated to anyone who could read. Movable type granted the text an authority that had previously belonged solely to the church or the monarchy. The coming centuries would witness a series of bloody struggles by which competing social groups tried to claim that authority for themselves.

Left-Brained Lutheranism

In fifteenth-century Germany, as printers started turning out mass quantities of Bibles and works by classical authors, they also started producing pamphlets, handbills, posters, and broadsheets (the forebears of the modern newspaper). Europe was entering a new era of mass communication, fueled by a disruptive information technology whose effects no one could possibly have foreseen. On October 31, 1517, Luther catalyzed the revolution of mass literacy when he nailed his ninety-five theses on the door of the Augustinian chapel in Wittenberg.

Without the printing press, there likely would have been no Protestant movement. The ease of printing and speed of communication allowed the written word to circulate freely outside the traditional institutional control valves of the church. Before the advent of the printing press, Rome could quash heretics at will; now the "heretics" had a powerful technology at their disposal. Hurriedly printed posters like Luther's soon became the core communications vehicle for the reformist movements that were beginning to percolate across Europe. Whenever reformists wanted to organize a public meeting, they would draw up a poster, have it printed, and tack it up all over town. The networked communications platform of the posted broadsheet would forever destabilize the old hierarchy of the Roman church.

"Lutheranism was from the first the child of the printed book," writes the historian Elizabeth Eisenstein, "and through this vehicle Luther was able to make exact, standardized and ineradicable impressions on the minds of Europe."[23] Luther hailed printing as "God's highest and extremest act of grace, whereby the business of the Gospel is driven forward."[24] If Christianity is the religion of the book, Protestantism is the denomination of the press.

Protestantism represented more than just a doctrinal departure; it signaled a psychic rift between two very different ways of understanding the world. As

Leonard Shlain puts it, "Luther repudiated the colorful ritual of mass, rich with icons. The nexus of the Reformation was the rekindling of the age-old conflict between written words and images."[25] The technology of movable type privileges the written word over the image, elevates left brain over right. Lutheranism represented a cognitive schism as much as a religious one. And while the social and political causes of the Reformation are too complex to catalog in full here, the printing press played a central role in triggering a series of bloody convulsions that would sweep the European continent.

In our modern literate culture, we tend to view the printing press as a force of liberation and an unqualified historical good, the technological wellspring of modern culture. At the time, however, the abrupt introduction of the new technology wreaked a series of calamitous social disruptions. The popular heroic image of Luther nailing his theses to the wall tends to divert our attention from the brutal truth of the Reformation. The Roman church made every effort to suppress the new polemic literature. The church issued a papal bull against Luther's work (which Luther promptly set aflame in protest). When Luther's work started attracting a following in Paris, the church issued another bull forbidding printers from publishing not only Luther but any work not bearing the church-sanctioned imprimatur of the University of Paris. In the wake of the papal ban, a black market soon sprang up as foreign printers recognized the opportunity to supply Parisian readers hungry for forbidden works. In an effort to stanch the flow of radical and potentially destabilizing books, the king of France issued an edict banning the import of all foreign books. Printers freely ignored the law. Meanwhile, sales of thoughtful writers like Erasmus began to decline in favor of the strident populism of Luther and his ilk. "As Luther's message caught on," writes Shlain, "the illiterate masses nonetheless revolted against the image-centric Church artifacts: smashing statues, slashing paintings, overturning altars. They built new Protestant churches with plain white walls, no images or stained glass—not even a crucifix. These were strictly left-brained affairs dedicated to interpreting the Word—not the Image—of God."[26]

In Germany, Luther inaugurated a stream of religious polemics that took on increasingly poisonous tones. Popular pamphlets with titles like "Pope Donkey" and "Cow Monk" painted vicious satires of the Roman clergy, inciting the lay population to acts of horrific brutality. During the battles of the Reformation, Europe descended into a period of atrocities in which opposing bands of Protestants and Catholics literally gutted each other in the streets. In Holland, when Dutch Calvinists invoked Protestant doctrine as their basis for overthrowing the Spanish Catholic government, the struggle boiled over into a frenzy of brutality. According to the writer John Lothrop Motley, "On more than one occasion, men were seen assisting to hang with their own hands and in cold blood their own brothers, who had been taken prisoners in the enemy's ranks. When the captives were too many to be hanged, they were tied back to back, two and two, and thus hurled into the sea. . . . On one occasion, a surgeon at Veer cut the heart from a Spanish prisoner, nailed it on a vessel's prow, and invited the townsmen to come and fasten their teeth in it, which many did with savage satisfaction."[27]

Wherever the Protestant message spread, powered by the new technology of printing, violent upheavals ensued. The century after Gutenberg witnessed a series of violent upheavals that raged across the continent. Peasants embraced the newly translated vernacular Bibles with a revolutionary fervor. Across Germany, tens of thousands of peasants refused to pay taxes. Peasant wars broke out, leaving over one hundred thousand dead. Of course, much of the violence was orchestrated by Catholics as well, inspired by their own understanding of a theology that is, at least in some sense, a "product" of the book. "Despite having recently read the New Testament for the first time," writes Shlain, "many people on both sides behaved in a manner antithetical to the spirit of the Gospels."[28]

Wherever the printing press took hold, conflict seemed to follow. Shlain argues that the introduction of printed books seems to have triggered a kind of mass social pathology that may have stemmed from the jarring introduction of a linear left-brain communications mode of thought into what had previously been a predominantly oral and visual right-brain culture. Shlain believes this cognitive disruption explains why printing and literacy seem to have spread in almost perfect lockstep with the rise of witch-burning. As printers brought the left-brained modality of the written word into contact with the old right-brain values of folk traditions and feminine-centered wisdom (as exemplified in the flourishing cult of Mary), masculine values came into open and violent conflict with feminine values. Countless women paid a tragic price. "It was in Germany," as Shlain points out, "the birthplace of the printing press and home to the fastest rise in literacy rates, that the witch craze assumed its most monstrous proportions."[29] In one town, 133 witches were executed in one day. In the two villages surrounding Trier, the witch hunters were so successful that only one woman was left alive. Between 1623 and 1631, the bishop of Würzburg (home of the great monastic library) executed forty-one girls ages seven to eleven.

The sudden shift from an oral, visually symbolic culture to an increasingly left-brained world of linear written texts may have triggered a deep shift in the European psyche that led, for a time, to a kind of mass psychosis. Paradoxically, when the book first emerged in the latter days of the Roman Empire, it had represented a new, more open kind of technology—a kind of analog hypertext— that freed readers from the unidirectional constraints of ancient scrolls. The days of the Irish scribe perusing classical texts were long past, however. Now books were turning into commodities, and the insistent logic of automated production gave them an increasingly linear character. Page numbers appeared, while illustrations dwindled. Unlike the handmade, carefully illustrated manuscripts of the monasteries, the printed book seemed to insist on forward progress.

As the printing press took root amid a series of larger social, political, and religious transformations, newly literate Europe started acting out. "The eerie conjunction of the printing press, steeply rising literacy rates, religious wars and the witch craze seems significant."[30] The sudden rise of alphabetic literacy seems to have accompanied a deep conflict in the collective social brain: a war between left and right brain values, in which left-brain values, for a time, emerged victorious after a violent struggle.

Today, many cultural observers decry the fall of civilized discourse in the age of social media and 24/7 cable news. But with the benefit of a deeper historical perspective, we might take solace in knowing that this is hardly the first time a new and untested mass communications medium has unleashed a period of partisan stridency in the public square. The explosion of printing wreaked havoc with the old institutional hierarchies. Europe was headed into a period of bloody upheaval. Those disruptions echoed in the more pacific world of scholarship as well. No longer could books be easily classified into the simple old Vatican systems of secular and sacred knowledge. There were simply too many books; moreover, an explosion of facts and printed knowledge was taxing the underlying ontologies of knowledge that had guided Western European understanding of the phenomenal world. Disciplinary boundaries were blurring, and the sheer volume of knowledge demanded a more methodical system for organizing humanity's burgeoning intellectual output. The old monastic scriptoria would soon die out; but as the era of printing dawned, the old scholastic ways had not quite breathed their last.

THE ASTRAL POWER STATION

Yea, from the table of my memory
I'll wipe away all trivial fond records,
All saws of books, all forms, all pressures past
That youth and observation copied there
And thy commandment all alone shall live
Within the book and volume of my brain.

—William Shakespeare, *Hamlet*

In 1532, an Italian fellow named Viglius wrote to his friend Erasmus about his upcoming trip to Venice. He was looking forward to seeing a new invention that had been causing a stir among the town's cognoscenti: a so-called theater of memory. Viglius was not entirely sure what the thing was, but he had heard that it was "a work of wonderful skill, into which whoever is admitted as spectator will be able to discourse on any subject no less fluently than Cicero."

Returning from Venice a few weeks later, Viglius wrote again to tell his friend what he had seen. "The work is of wood," he wrote, "marked with many images, and full of little boxes." The theater turned out to be a contraption the size of a small room, housing a complicated apparatus with tiny gears that opened and closed windows to reveal words and images printed inside. By manipulating the windows, a visitor could retrieve nuggets of information on any number of topics: the seven virtues and vices, the teachings of Solomon, the orientation of celestial bodies, and all manner of other medieval factoids. It was a kind of sixteenth-century Rube Goldberg hypertext machine. Step inside, click open a window or two, and presto: the wisdom of the ancients revealed. Viglius reported that the theater's inventor, Giulio Camillo, "calls this theatre of his by many names, saying now that it is a built or constructed mind and soul, and now that it is a windowed one."

For the newly literate Venetian audience, Camillo's contraption must have seemed nothing short of a technological marvel. It was also a perfect metaphor for a dawning era of post-Renaissance information technology. The spread of popular literacy and the erosion of monastic power had created a fertile environment for a new breed of maverick scholars to emerge. Freed of the intellectual confines of the monastic system, a few pioneering thinkers began to experiment with new ways of structuring access to human knowledge: from encyclopedic contraptions like Camillo's to mystical memory wheels, universal classification schemes, and artificial languages. Many of these ambitious efforts would meet unfortunate ends, with their progenitors ending up in bankruptcy, burned at the stake, or the butt of famous literary jokes. But ultimately, the Renaissance information technologists paved the way for the emergence of the scientific method and a new secular approach to information systems that still reverberates today.

No one knows exactly what Camillo's theater looked like—no drawings have survived—but Camillo once described the design as follows: "The Theatre rests basically upon seven pillars, the seven pillars of Solomon's House of Wisdom. Solomon in the ninth chapter of Proverbs says that wisdom has built herself a house and has founded it on seven pillars. By these columns, signifying most stable eternity, we are to understand the seven Sephiroth of the super-celestial world, which are the seven measures of the fabric of the celestial and inferior worlds, in which are contained the Ideas of all things in the celestial and in the inferior worlds."[1]

The machine worked by allowing users to tunnel their way through a series of nested conceptual hierarchies, proceeding from the gross physical plane of the theater to the metaphorical House of Wisdom, then on through successive layers of abstraction, finally ascending to the realm of divine truths to pierce the innermost secrets of the arts. This grand scheme was not Camillo's own. It actually represented a cheap facsimile of an ancient scholastic practice that had flourished in Europe's medieval monasteries for centuries: the art of memory. Camillo, himself a former monk, had left the cloisters several years earlier to seek his fortune in the town squares of Europe. Determined to make something of himself, he hit on the idea of taking a few tricks he had learned in the monastery and turning them into a public attraction. He even came up with a catchy tagline: turning "scholars into spectators."

The original monastic art of memory was a far cry from Camillo's version. In medieval monasteries, generations of monks had perfected a demanding mental regimen that enabled them to memorize long tracts of scripture and other information through years of difficult training. The technique traced its roots to the Greek poet Simonides of Ceos, who in the fifth century BC described a method for improving one's memory by visualizing a series of loci (places) in a particular order, then associating a meaningful image with each place, as an aid to reminiscence. Aristotle later picked up on Simonides's work, incorporating the technique into his discussion of memory in his *De memoria et reminiscentia* (On memory and recollection) in which he described the act of memorization as a visual practice that invoked the "inner eye" of memory to summon *phantasmata* (images) stored in the "sensitive soul" and making them available to the "intellectual soul" of the logical mind. Aristotle's description of the technique eventually found its way to Rome, where Cicero, Quintilian, and other rhetoricians embraced the art of memory as an aid to practicing the great Roman art of oratory, using the technique to command vast arrays of facts, quotations, ideas, and topics that they could pull together extemporaneously while delivering orations.[2]

The classical art of memory faded with the decline of Rome, and Aristotle's seminal work on the subject fell into obscurity until the thirteenth century, when European scholars rediscovered his work *On Memory* (*De memoria et reminiscentia*). Thomas Aquinas, in particular, wrote a series of commentaries on Aristotle's work that helped popularize the technique among medieval scholastics, celebrating the technique as a remedy for the inherent frailties of the human mind. Aquinas pointed out that although the mind was well equipped to recollect the gross details of physical experience, it was ill suited for remembering "subtle and spiritual things" (a limitation stemming from man's fallen condition). The art of memory, as he saw it, offered monks a powerful tool for harnessing their physical and spatial memory as a means for internalizing God's word, invoking Aristotle's assertion that "the soul never thinks without a mental picture."[3] Aquinas went on to describe the art as relying on two essential facilities: association (the ability to relate one concept to another) and order (the ability to arrange spatial memories into nested categories)—in other words, networks and hierarchies.[4]

In its medieval incarnation, the art became a spiritual regimen, requiring years of contemplation and devotion in the quiet of a monastery. Cloistered monks, with no shortage of time on their hands, passed their days memorizing scriptures, religious commentaries, and assorted classical works. At first, monastic practitioners seem to have viewed the art as a kind of practical information technology. One monk, John Willis, postulated that memory "so far as it is strictly taken for the common receptacle of Memorandums, is merely passive . . . in the same manner as Paper preserveth words written therein." He goes on to caution, however, that just as "it is the office of a Scribe, not of Paper, to write and read things written; so to dispose Idea's in Memory, and aptly to use them, is work of Understanding, not of Memory."[5] Memorization, in other words, is no substitute for true learning.

The monastic version of the technique worked something like this. A monk would apprentice himself to a master of the art, who would teach him to visualize a series of metaphorical images. These might include buildings, palaces, cities, wax tablets, chains or writing. For example, a monk might learn to memorize a room in a house, containing a series of symbolic objects tied to a particular theme. The theology room might house (surprise) a theologian, his head tattooed with the words *cognitio, amor, fruitio*; his limbs with the words *essentia divina, actus, forma, relatio, articula, precepta, sacramenta*; and so on.[6] The trainee would commit the contents of each room to memory, slowly mastering the painstaking journey from room to room.

Despite the volume of published writing on such a highly visual technique, illustrated depictions are surprisingly rare. The art was largely handed down through direct oral transmission, without external visual aids. One of the few tangible glimpses survives in a Florentine convent, where a fresco depicts Thomas Aquinas seated on a throne amid flying figures representing the seven virtues; to each side sit the saints and patriarchs; beneath his feet lie vanquished heretics; farther down, seven female figures represent the seven liberal arts; then another seven represent the theological disciplines.[7] The fresco provides one of the few physical manifestations of a technique that otherwise resided, and expired, only within its practitioners' minds.

The practice took years to master, but those who succeeded reported astonishing mnemonic feats. One practitioner, Peter of Ravenna, claimed to have memorized more than one hundred thousand rooms, enabling him to recall from memory the entire canon law, including every related commentary, the full text of two hundred complete speeches by Cicero, three hundred philosophical sayings, twenty thousand legal points, and more.[8] Another enthusiast, Johannes Romberch, in 1533 published a widely read grammar that laid out an elaborate system for memorizing the entirety of theology, metaphysics, law, astronomy, geometry, arithmetic, music, logic, rhetoric, and grammar.

From the late fifteenth century onward, numerous scholastic writers penned treatises on the technique, including Giovanni de Carrara's *De omnibus ingeniis augendae memoriae* (1481), Peter of Ravenna's *Phoenix, sive artificiosa memorabilia*, Guglielmo Gratoroli's *De memoria* (1553), Cosimo Rosselli's *Thesaurus artificiosae memoriae* (1579), and William Fulwood's *The Castle of Memorie* (1562). As European scholastics embraced the art in growing numbers, the practice became increasingly esoteric, eventually devolving into what the historian Frances A. Yates calls "a kind of cross-word puzzle to beguile the long hours in the cloister." Often, the monks found themselves turning these memory puzzles into intricate and time-consuming diversions that "can have had no practical utility; letters and images are turning into childish games."[9] Memorization became a kind of monastic fad, often at the expense of learning.

So Camillo's theater represented the culmination of an apostasy long in the making, the conflation of memorization with learning. With the decline of monasticism, however, it also represented an important effort to translate the previously secretive technique for a broader lay audience. And although the simplistic literalism of Camillo's boxes and windows may have represented a debased version of the original art, it also marked a philosophical turning point at the beginning of the Renaissance. It represented, as Yates puts it, "a new Renaissance plan of the psyche, a change which has happened within memory, whence outward changes derived their impetus." By giving the once-ethereal art a concrete form, Camillo transformed the old monastic art into a vehicle for secular aspiration. In so doing, he also anticipated the broader philosophical trajectory of the Renaissance. For the medieval monastics, the art of memory was an antidote to the weakness of the human mind; by contrast, "Renaissance Hermetic man believes that he has divine powers; he can form a magic memory through which he grasps the world, reflecting the divine macrocosm in which the microcosm of his divine *mens*."[10]

Camillo, it should be noted, never actually completed his theater. He died before he could finish raising the funds to complete his ambitious vision. Soon after Camillo's death, the first generation of European humanists began to decry the old scholastic ways. Writers like Erasmus (Viglius's friend) criticized the old medieval art, enjoining scholars to embrace a new rationalism that relied on more "natural" or linear forms of memory. Other influential European thinkers like Philip Melanchthon, Cornelius Agrippa, and Michel de Montaigne all pilloried the monastic art of memory. Thomas Fuller bemoaned the "Memory-mountebanks" who had corrupted the classical art of memory by locking it in the cloister and keeping it from being used for practical purposes.[11] "For the Erasmian type of humanist the art of memory was dying out," writes Yates, "killed by the printed book, unfashionable because of its mediaeval associations, a cumbrous art which modern educators are dropping."[12] For the early modern humanists, the art of memory "belonged to the ages of barbarism; its methods in decay were an example of those cobwebs in monkish minds which new brooms must sweep away."[13]

Bruno's Heresy

As Europe entered into a new period of secular reason and scientific inquiry, the old art of memory would slowly fall into disrepute. At the end of the sixteenth century, however, it would see one more burst of popular interest, thanks in part to the high-profile heresies of Europe's great intellectual maverick: Giordano Bruno.

On February 9, 1600, Bruno bowed on one knee before the grand inquisitor, after six years in papal custody. The charge was heresy, and the inquisition had pronounced him guilty. When the executioners presented Bruno with a cross at the moment of his death, he waved it away. He thought he had already found his salvation.

Born in 1548, Bruno had trained as a Dominican friar in Naples, where he had studied and mastered the art of memory. He eventually forsook the monastery for a wandering life that took him through France, England, and Germany. Along the way, he began teaching the art to interested lay students, including no less a personage than King Henri III of France. Bruno would later recall that when the king summoned him to court, he first "asked whether the memory which I had and which I taught was a natural memory or obtained by magic art; I proved to him that it was not obtained by magic art but by science."[14]

In truth, however, Bruno's version of the art contained a great deal of "magic." Bruno saw the practice not just as a tool for strengthening the memory but as a transcendental gateway to "innumerable worlds." He expanded the scope of the art beyond the usual doctrinaire topics of the monasteries, using it to integrate the classical wisdom of the Greeks, the magical religion of the Egyptians, ultimately penetrating, he believed, the highest truths of creation. In Bruno's rendition, the art joins the lower world of physical reality with the incorporeal essences of the celestial sphere, for "it is by one and the same ladder that nature descends to the production of things and the intellect ascends to the knowledge of them."[15] In attempting to join heaven and earth, Bruno embraced an Aristotelian conceit: that the phenomena of the material world are expressions of inner essences, that things are, in a sense, symbols of themselves. It was, as Yates puts it, "a highly systematized magic."[16]

Bruno interpolated the old art into a new occult form of practice, claiming to have unlocked the deepest mysteries of the phenomenal world thanks to a complex mnemonic device called the memory wheel. The memory wheel relied on a series of ostensibly magic numbers, especially 30 and 150, with 150 images broken out into five sets of thirty images each, all arranged in concentric circles, or wheels.[17]

In a meticulous feat of literary reverse engineering, Yates managed to decipher Bruno's dense text in sufficient detail to construct a working model of the memory wheel—a feat that, as far as we know, no one had attempted in four hundred years. She describes how she believed the mnemonic device must have operated:

> The lists of images given in the book are marked off in thirty divisions marked with these letters, each division having five subdivisions marked with the five vowels. These lists, each of 150 images, are therefore intended to be set out on the concentric revolving wheels. Which is what I have done on the plan, by writing out the lists of images on concentric wheels divided into thirty segments with five subdivisions in each.

FIGURE 20. Memory wheel, from Raimundus Lullus's *Opera*, 1617.

The result is the ancient Egyptian looking object, evidently highly magical, for the images on the central wheel are the images of the decans of the zodiac, images of the planets, images of the mansions of the moon, and images of the houses of the horoscope. The descriptions of these images are written out from Bruno's text on the central wheel of the plan. This heavily inscribed central wheel is the astral power station, as it were, which works the whole system.[18]

The system builds on wheel after concentric wheel, spinning out into innumerable combinations whose mastery demands extraordinary dedication and concentration. Bruno's path calls to mind the tantric practices of Tibetan Buddhism, which similarly rely on intricate visualizations and extreme feats of memorization as pathways to spiritual liberation. Bruno's "astral power station" was a prodigious piece of conceptual engineering, with its endless combinations of juxtaposed letters revealing layer upon layer of nested meaning. At the first level, the wheel contains an inventory of empirical facts concerning the kingdoms of animals, vegetables, and minerals. Further turnings of the wheel reveal new layers of abstraction, as the concentric circles expand to catalog the whole sweep of human achievement, encoding the names of mythical figures who represent the enabling technologies of civilization: Pyrodes, the inventor of fire; Aristeus, the discoverer of honey; Coraebus, the potter; and so on. The list then moves through more advanced technologies like glass blowing, surgery, writing (represented by our old platonic friend Thoth), and up through the rarefied magical arts like necromancy, chiromancy, and pyromancy, and then to higher and higher degrees of celestial wisdom. It is a breathtakingly complete array of the known intellectual universe. And with a touch of authorial chutzpah, Bruno places himself squarely in his own cosmology, invoking himself as Ior (short for Iordano Bruno), the inventor of the "key."

Whereas the medievalists of the Thomist tradition treated the art of memory as a remedy for the failings of the human mind, Bruno saw it as a fruitional path, a means to put people in touch with their innate divinity. "If you contemplate this attentively," Bruno wrote, "you will be able to reach such a figurative art that it will help not only the memory but also all the powers of the soul in a wonderful manner."[19] Like the Gnostics, Bruno thought he had found a path of liberation that relied purely on training the mind, rather than on pious obedience to the top-down dictates of the church.

Although Bruno based his memory technique on the art of memory that would soon fall into disfavor among Renaissance scholars, his particular use of the technique anticipated the scientific method. By using the technique to articulate a system for classifying information about the natural and cosmological worlds and articulating the rules of that system in writing, he paved the way for the emergence of empirical methods that would soon transform Renaissance scholarship.

In Howard Bloom's parlance, Bruno would qualify as a first-rate diversity generator, a fount of alternative hypotheses who failed to survive the intergroup tournament—the inquisition—and ultimately perished at the hands of conformity enforcers—the church (see chapter 1).[20] But despite Bruno's unfortunate demise, his ideas found their way into the mainstream of Renaissance thought, paving the way for the advent of the scientific method.

Ultimately, Bruno's practices amount to an aberration from the art of memory as it had largely been practiced beforehand. As the historian Rhodri Lewis writes, "Bruno's ambitions—if not his achievements—were at variance with the history of mnemotechnique."[21] Throughout most of its history, the art of memory had been a practical tool for improving recollection; it was only in the latter years of the scholastic fascination that it began taking on mystical airs. After Bruno's execution, the monastic practice of the art of memory faded almost entirely from public view, but traces of its impact would linger for centuries to come. An Englishman named Robert Fludd wrote about the practice in the seventeenth century. Some scholars have even speculated that the art of memory went back underground into esoteric occult circles, as it had during the Dark Ages. Yates believes that Bruno's writings also embedded themselves into the secret practices of the Freemasons and Rosicrucians. Esoterica aside, however, the old medieval knowledge systems—with their cathedrals of imagery, theaters of memory, and books of mythological animals—were fast giving way to a new era of scientific reason. Soon, the old monkish ways would be all but forgotten, thanks in no small part to an Englishman well versed in the art of memory: Francis Bacon.

Bacon's Window

The most striking feature of Francis Bacon's house in Gorhambury, England was a lattice of stained-glass windows, "every pane with severall [sic] figures of beast, bird and flower."[22] Bacon, who had studied the old art of memory in his youth, had taken great pains to create a physical facsimile of the old scholastic art in his house. Perhaps we can envision him gazing out his multicolored window, imagining the memory palaces he had visited in his monastic youth.

By the time Elizabeth I ascended to the throne in 1558, English intellectuals had started abandoning the mystery cults of the scholastics. The Cambridge-educated Bacon ultimately came to deride the old scholastic methods as overly stylized exercises in meaningless abstraction. His years in the monastery had given him a firsthand perspective on the futility of esoteric debates that seemed removed from practical problems: the medieval equivalent of the ivory tower syndrome. Along with Erasmus and other early humanist writers, he condemned the art of memory—in which he himself had trained—as a superfluous exercise that paled in comparison to an empirical style of reasoning based on direct observation and logical reasoning. "We are certain that there may be had both better Precepts for the confirming and increasing *Memory*, than that Art comprehendeth," he wrote, "and a better Practice of that Art may be set downe than that which is receiv'd." "It seemeth to me that there are better practices of that art than those received," he wrote of the art of memory, deriding the monastic tendency to celebrate their elaborate memory feats as "points of ostentation."[23] He further derided the practitioners of the monastic art as a kind of intellectual gymnastics practiced by "Tumblers; Buffones, & Iug[g]lers."

Despite his condemnation of the monastic art of memory, Bacon did not reject the technique altogether but yearned for a "better Practice of that Art," stripped of scholastic excess. In his landmark *Novum organon*, he recommended the use of "artificial places in memory, which can either be places in the literal sense . . . or familiar or famous persons, or whatever you like (provided that they are organized properly)."[24] Bacon longed for a new intellectual philosophy that rejected the cloistered model of scholastic learning in favor of practical inquiry into real-world problems. Although he rejected the idle contemplations of the monastics, he nonetheless recognized the value of mnemonic techniques as a bridge toward a new way of understanding the natural world.

Although history has remembered Bacon as one of the forebears of the scientific method, he did not envision himself as a "scientist" (the term did not yet exist). Rather, Bacon thought of himself as a natural philosopher, or even as a "magician." In the medieval sense of the term, a magician was not a conjurer or sleight-of-hand artist but an explorer of the phenomenal world, devoted to exploiting the laws of nature for practical use. Bacon practiced alchemy and saw no conflict between practicing magic and looking for empirical methods. "Toward the effecting of works," he wrote, "all that man can do is put together or part asunder natural bodies. The rest is done by Nature working within."[25] In this sense, magic is just another word for science. Elsewhere, Bacon wrote that "the aim of magic is to recall natural philosophy from the vanity of speculations to the importance of experiments." The duty of a philosopher, as Bacon saw it, was not to reject magical traditions out of hand but to delve into them with fresh eyes and a critical perspective. "For although such things lie buried deep beneath a mass of falsehood and fable, yet they should be looked into . . . for it may be that in some of them some natural operation lies at the bottom."[26]

By invoking the power of observation to sift nature's truth from the "vanity of speculations," Bacon forged a new path of understanding the natural world, based not on faith in a divine truth but on the power of individual perception. He rejected monastic scholasticism as a form of mysticism that placed undue faith in divine revelation, preferring instead to trust the individual's powers of observation; Bacon belonged squarely to the age of reason. In 1620, Bacon introduced his most famous formulation of this new way of knowing in his greatest work, the *Novum organum* (New tool of reason).[27] Here, he proposed nothing less than restructuring the enterprise of scholarship. The book laid the philosophical foundations for a process that would later become known as the scientific method, a radical departure from the scholasticism that sought pathways to truth through esoteric practices and belief in disembodied ideals. Bacon believed in the capacity of direct perceptual observation to reveal God's truth, rejecting appeals to external authorities like classical authors or even the scriptures. "I have taken all knowledge to be my province," he wrote—with no small display of authorial hubris—staking out an ambitious philosophical agenda in the form of a six-part approach to investigating the natural world.

The title page opens with a grand illustration depicting the words of the title rising above two ships sailing between the pillars of Hercules, symbolizing the limits of human knowledge in the ancient world. By invoking "the pillars of fate set in the path of knowledge," Bacon signaled his ambition to explore an entirely new kind of knowledge. The book is an intellectual tour de force, a sweeping foray into a new mode of reasoning that would ultimately fulfill Bacon's ambition and reshape the trajectory of Western scholarship. Bacon proposed a new method of reason based on induction—reasoning from direct observation—and relying on an idealized system of scholarly collaboration that viewed knowledge as a cumulative enterprise. In other words, he favored networked communication between scholars over received top-down hierarchical modes of understanding, be they classical or ecclesiastical. "I do not endeavor either by triumphs of confutation, or pleadings of

FIGURE 21. Cover plate of Bacon's *Instauratio magna* (1620).

antiquity, or assumption of authority, or even by veil of obscurity, to invest these inventions of mine with any majesty," he wrote. Instead, he aspired to "present these things naked and open, that my errors can be marked and set aside before the mass of knowledge be further infected by them; and it will be easy also for others to continue and carry on my labours."[28] His vision of research as a collaborative process, subject to trial and error—rather than as the revelation of divine truths— marks the earliest expression of what we now recognize as the scientific method.

For all his belief in collaborative networks of scholarship, however, Bacon was no populist. He frequently voiced undisguised contempt for the common man, believing that the great uneducated masses posed a severe threat to the integrity of scholarship. Paradoxically, Bacon at once advocated the democratization of scholarship while espousing intellectual elitism. He believed that in order to achieve trustworthy results, scholars had to work to overcome the inherent human limitations that keep the majority of the population in the figurative dark. Bacon described these barriers to understanding as "idols," which he described as "the profoundest fallacies of the mind of Man" that spring "from a corrupt and crookedly-set predisposition of the mind. . . . For the mind of man . . . is rather like an enchanted glass, full of superstitions, apparitions, and impostures.[29] Bacon's idols include the following:

- *Idols of the Cave.* The problem of subjectivity. Each of us inhabits "a cave or den of his own, which refracts and discolors the light of nature." Our personal biases will always taint our perceptions. We might today say that the chemist sees chemistry in all things; the student of literature looks for narratives; the computer programmer looks for operable rules. Relying on our individual faculties alone can only lead us to imperfect, idiosyncratic versions of reality.
- *Idols of the Tribe.* The problem of human limitations. These are the inborn perceptual constraints of our human "tribe." "Human understanding is like a false mirror, which, receiving rays irregularly, distorts and discolors the nature of things by mingling its own nature with it." We cannot, for example, perceive the full spectrum of colors and sounds; but through the careful application of logical reasoning, we can begin to compensate for our sensory limitations.
- *Idols of the Marketplace.* The problem of socially constructed meaning. Our understanding will always be warped by the vagaries of human language. "It is by discourse that men associate, and words are imposed according to the apprehension of the vulgar. And therefore the ill and unfit choice of words wonderfully obstructs the understanding." Words are imperfect vessels, pale approximations of direct experience.

- *Idols of the Theater.* The problem of belief. Mythologies, religious stories, or ideological convictions present enormous obstacles to understanding, like "so many stage plays, representing worlds of their own creation after an unreal and scenic fashion." Only by renouncing received wisdom and superstitious beliefs, and refusing to take anything on faith, can we cultivate any hope of seeing things as they are.

Bacon's idols constitute a sweeping rebuke of the old medieval mindset, predicated as it was on superstitious belief in the word of God or the classical ideal of truth. Bacon argued that the pursuit of scientific truth must be derived from direct observation of phenomena and a process he famously dubbed induction, which requires meticulous observation and logical reasoning. In contrast to Aristotle's top-down system of ideal forms stemming from disembodied conceptual hierarchies (his great chain of being), Bacon proposed a radical new system through which hierarchies of meaning emerged from the bottom up, through the collective effort of scholars creating conceptual building blocks of observed phenomena.

Bacon took an active and hopeful view of human agency, believing that the pursuit of empirical knowledge was a vehicle for human salvation. "Let us hope," he wrote, "that there may spring helps to man, and a line and race of inventions that may in some degree subdue and overcome the necessities and miseries of humanity."[30] Ultimately, Bacon's guiding philosophy was a steadfast conviction that "useful knowledge," rather than the idealistic pursuit of truth, would lead to humanity's salvation.

Bacon's pursuit of an empirical method would eventually lead him to formulate a new philosophical framework for classifying all of human knowledge. He postulated that all human intellectual pursuits revolve around three essential facilities: memory, reason, and imagination (or, in more familiar terms, history, philosophy, and poetry). In *De augmentis scientiarium* (1623), he proposed a new scheme for organizing human knowledge. Echoing Richard de Fournival's early classification scheme for the Sorbonne library, Bacon made a top-line distinction between divine and human learning, with the latter category branching into history, poetry, and philosophy (encompassing the natural sciences). Even as he proposed a universal classification scheme to govern human knowledge, at the same time he acknowledged the fundamental shortcomings of any attempt to draw clear lines between disciplines, and the importance of cross-disciplinary inquiry. He hoped that "all partitions of knowledge be accepted rather for lines and veins, than for sections and separations."[31] Nonetheless, this simple model would also prove to be one of Bacon's enduring legacies, influencing the thinking of information scientists for centuries to come. As we will see in later chapters, Denis Diderot would embrace Bacon's scheme as the foundation for his great encyclopedia; Thomas Jefferson would later adopt it as the basis for organizing his own personal library, the foundation for the Library of Congress; and Melvil Dewey would acknowledge it as an influence on his decimal system.

For all of Bacon's subsequent historical impact, in his day his work met with a largely muted reception. His chief patron, King James I, appraised his work with a royal backhanded compliment, noting that he found his lord chancellor's work—like the peace of God—"passeth all understanding." Bacon acknowledged the density of his work, likening his undertaking to guiding readers through a "labyrinth." He never tried to sugarcoat the difficulty of finding a path to true understanding, an effort "presenting on every side so many ambiguities of way, such deceitful resemblances of objects and signs, natures so irregular in their lines, and so knotted and entangled," that the steps of the true method "must be guided by a clue," which Bacon envisioned as "a more perfect use and application of the human mind and intellect."[32]

Wilkins's New World Order

On January 11, 1666, the great English diarist Samuel Pepys wrote of meeting with the well-known London bishop Dr. John Wilkins. Pepys had heard that Wilkins was working on an ambitious scheme to develop a "Universall Language," a new way of cataloging human knowledge. Wilkins had invited Pepys to contribute his knowledge of naval matters to the project, and Pepys eagerly answered the call. So too did many other prominent members of the newfound Royal Society, under whose auspices Wilkins had undertaken his project. John Ray assisted "in framing his Tables of Plants, Quadrupeds, Birds, Fishes &c. for the User of the universal Character." Other notable collaborators included the botanist Robert Morison, the naturalist Francis Willughby, and the clergyman William Lloyd.[33] Although the system would ultimately bear Wilkins's name, it was largely a collaborative effort.

In 1668, Wilkins published his landmark *Essay toward a Real Character and a Philosophical Language*. Weighing in at a prodigious 454 folio pages, the *Essay* is divided into four sections: an introduction, the tables of "Universal Philosophy" that comprise the bulk of the work, the grammar of a new philosophical language, and finally a section explicating how the character and language work together. In the Universal Philosophy, Wilkins proposes classifying the phenomenal world into forty basic categories, beginning with two top-level divisions— "General or universal notions" and "Special" things (see appendix A for a complete list of Wilkins's categories). These categories are then subcategorized by "differences" and further by individual "species," for a total of 4,194 possible classifications. The system works by generating four-letter names based on the rules of the classification: the first two letters corresponding to the genus, the third letter (always a consonant) identifying the difference, and the fourth letter (always a vowel) identifying the species. For example, the genus *Zi* signifies "beasts"; within which the difference *p* refers to "rapacious dog-like beasts" (including, for example, both wolves and dogs); within which the final signifier *a* identifies the particular species "dog." Thus, *zipa* means "dog."

In an age of burgeoning secular scholarship, Wilkins believed the world desperately needed a new synthetic language capable of containing the growing surfeit of available data. He thought that the world's old natural languages had degenerated, becoming "inartificial and confused," and were poorly suited to accommodate the truth of God's creation. Older languages, he believed, were fundamentally flawed because of their haphazard development, by "that Curse in the Confusion of Tongues" that left them bedeviled with redundant and imprecise meanings. Like Bacon, Wilkins believed that human comprehension was limited by our preconceived notions about things (cf. Bacon's "Idols of the Marketplace"). To remedy the shortcomings of written languages, he set out to formulate a new system capable of describing such "natural truths as are to be found out by our own industry and experience," untainted by existing linguistic biases. As the Wilkins scholar Rhodri Lewis puts it, "Wilkins believed that it was possible accurately to map the order of thought (and therefore of the things which this represented), and that it was possible to devise a language which might represent this."[34] To that end, Wilkins set out to create a language "that should not signifie *words*, but *things* and *notions*." His classification, he hoped, would provide "a sufficient enumeration of all such things and notions, as are to have names assigned to them, and withal so to contrive these according to their order, that the place of everything thing may contribute to a description of the nature of it."[35]

For all its exhaustive ambition, Wilkins's scheme explicitly excluded many possible topics, including fictional and "complex" phenomena as well as place names, clothing, games, food, music, and curiously, religious sects. He thought his language was best suited to describe simple, observable things and that more elusive concepts and ideas were best treated with "a more mixed and complicated signification."[36] Nonetheless, Wilkins believed that the truth of God's creation could be discovered primarily in observable nature, not through the received wisdom of the church. Elsewhere, he argued for the separation of scientific inquiry from religious doctrine, urging his contemporaries not to be "so superstitiously devoted to Antiquity as for to take up everything as canonical which drops from the pen of a Father."[37] Wilkins saw his life's work as an endeavor to bring humanity closer to God by forsaking superstitious beliefs in favor of honest intellectual striving.

For a time, it looked as though Wilkins's ambitions might be realized. The book met with an enthusiastic public reception after its initial publication. The

Royal Society began using the system to organize its collections, and copies of the book circulated throughout Europe. Wilkins himself hoped the *Essay* was only a beginning; for the language to reach its full realization, he thought, would require the efforts of "a College or an Age." "I am not so vain," he wrote, "as to think that I have here completely finished this great undertaking." He hoped that the *Essay* would serve as an ontological beacon for future generations.

Wilkins would doubtless have been dismayed to discover that centuries later, his life's great work is now best remembered not as the birth of a universal language but as the butt of a literary joke. Today, Wilkins is largely known as the target of Jorge Luis Borges's famous satire, "The Analytical Language of John Wilkins." Taking stock of Wilkins' system, Borges writes,

> Having defined Wilkins' procedure, we must examine a problem that is impossible or at least difficult to postpone: the value of the forty genera which are the basis of the language. Consider the eighth category, stones. Wilkins divides them into vulgar (flint, gravel, slate), middle-prized (marble, amber, coral), precious (pearl, opal), more transparent (amethyst, sapphire) and earthly concretions not dissolvible [*sic*] (pit-coal, oker and arsenic). Almost as surprising as the eighth, is the ninth category. This one reveals to us that metals can be imperfect (vermillion, mercury), factitious (bronze, brass), recrementitious (scoria, rust) and natural (gold, tin, copper). Whales belong to the sixteenth category; viviparous oblong fish.[38]

Borges went on to mock Wilkins's seemingly arbitrary classification with a modest taxonomic proposal of his own, masked as a fictional Chinese encyclopedia called *The Celestial Emporium of Benevolent Knowledge*. Here, he proposed a wildly orthogonal system for classifying animals as

a. belonging to the Emperor
b. embalmed
c. trained
d. sloppy
e. sirens
f. fabulous
g. stray dogs
h. included in this classification
i. trembling like crazy
j. innumerable
k. drawn with a very fine camelhair brush
l. et cetera

m. just broke the vase

n. from a distance look like flies

Borges concluded that Wilkins's system was doomed to fail because "it is clear that there is no classification of the Universe not being arbitrary and full of conjectures. The reason for this is very simple: we do not know what the universe is." Much as Borges's essay deserves its fame, we must ask whether it was perhaps a bit uncharitable to the unfortunate Mr. Wilkins. Wilkins never saw himself as any kind of absolutist; he railed against the arbitrary top-down hierarchies of the church and saw his system as, after a fashion, a bottom-up endeavor. He was an empiricist, not an authoritarian. Paradoxically, Wilkins might even have appreciated Borges's joke.

Beyond Monkish Minds

Barely a century after Camillo had wowed his Venetian audience with his gearbox theater of memory, Europe had witnessed a period of remarkable intellectual

FIGURE 22. Two Dutch scholars in the library of Leiden University, ca. 1610, from William Andrews, *Curiosities of the Church: Studies of Curious Customs, Services and Records* (1891).

fermentation. By the beginning of the eighteenth century, the old scholastic arts of the monasteries had faded to oblivion, supplanted by a new secular ethic of scholarship. The authority of the church had eroded to make way for a new mode of intellectual self-reliance best exemplified in the emerging scientific method and the dawning age of reason. Meanwhile, the ongoing caterwaul of publishing continued to fuel interest in a conceptual system capable of accommodating a growing tide of printed knowledge. The old wooden theaters, memory wheels, and artificial languages began to look like quaint relics of the bygone era of the monkish mind. Although none of these systems would last, these pioneering efforts mattered less for what they actually accomplished than for the vision they suggested. These brilliant failed experiments in mnemotechnology pointed the way toward an even greater undertaking that would occupy philosophers for centuries to come: the quest for a universal classification.

THE ENCYCLOPEDIC REVOLUTION

Until then I had thought each book spoke of the things, human or
divine, that lie outside books. Now I realized that not infrequently
books speak of books: it is as if they spoke among themselves.

—Umberto Eco, *The Name of the Rose*

With the rise of Renaissance secular scholarship and the emergence of the scientific method, seventeenth-century Europe witnessed a scholarly publishing boom. Across the growing university towns and printing centers of Europe, writers and printers turned out a growing volume of books on all kinds of topics: poetry, philosophy, politics, the arts, and the natural sciences, as well as traditional religious and devotional texts. Before 1500, European printers produced about fifteen million to twenty million books (from a total of 30,000–35,000 different titles); in the century that followed, that number mushroomed to nearly two hundred million (from 150,000 to 200,000 titles).[1] As books flooded the market, the burgeoning quantity of available information created the conditions for a new kind of book to emerge: the encyclopedia.

Today, we think of encyclopedias as dull reference works, secondary sources relegated to school libraries or bundled with our computers on CD-ROMs. In the seventeenth century, however, the encyclopedia represented a major innovation: a new literary technology that promised to help readers cope with a surging tide of new information. Encyclopedias did more than just give users an easy reference source; this new technology also had revolutionary implications.

In 1624, the English playwright Thomas Heywood published a widely circulated book called the *Gunaikeion*, a protoencyclopedia that purported to catalog all available information about his favorite subject: women. Inside this thick tome, readers would find tales of brave queens, learned ladies, chaste damsels, Amazons, witches, and even a transgendered woman or two. Although Heywood's name graced the cover, the *Gunaikeion* was a compilation of material culled from other popular books of the day, condensed, as Heywood put it, into "a small Manuell, containing all the pith and marrow of the greater."[2] Like a kind of Renaissance hypertext, the book had no table of contents, no conceptual "top" or Aristotelian subject hierarchy. Instead, it presented its contents in a loose structure that left the reader plenty of discretion in choosing how to navigate the text.

In synthesizing a vast body of recorded knowledge on the subject of women, Heywood grappled with what literary scholars Nonna Crook and Neil Rhodes call the "intractability of his subject matter." Rather than try to shepherd his fair subjects into a fixed classification, Heywood instead hit on a novel scheme: "I have not introduced them in order, neither Alphabetically, nor according to custome or president; which I thus excuse: The most cunning and curious Musick, is that which is made out of discords."[3]

The term *discords* seems like a useful rubric for the problem of organizing a large collection of information drawn from many disparate sources. In our era of distributed information systems, information architects struggle with the problem of imposing a workable order on an expanding body of information drawn from far-flung sources. Heywood anticipated this question long before anyone started using terms like *hypertext*, articulating a system for organizing information that did not invoke a top-down hierarchical classification but emerged through a dialogue between author and readers—a kind of ad hoc network—working together to interpolate the contents of other books that had come before. As Heywood put it, his new book was "all mine and none mine. As a good hous-wife out of divers fleeces weaves one peece of Cloath, a Bee gathers Wax and Hony out of many Flowers, and makes a new bundle of all. . . . I have laboriously collected this Cento out of divers Writers."[4]

For example, when Heywood recounts the startling deeds of a woman who wrote an illustrated sex manual, he wonders whether she might best be classified a "poetess" or a "she-monster." Similarly, when he tells of the great Queen Semiramis, a brave and noble woman of antiquity, he lets readers gauge whether she might best be tagged as a valorous queen, a murderess, a transvestite, or a practitioner of bestiality.[5] Writing of Queen Elizabeth, Heywood sings her praises in polymorphous terms: "Elizabeth, of late memory, Queene of England, she that was a Saba for her wisdome, an Harpalice for her magnanimitye . . . a Cleopatra for her bountie, a Camilla for her chastitie, an Amalasuntha for her temperance, a Zenobia for her learning and skill in language; of whose omniscience, pantarite, and goodnesse, all men heretofore have spoke too little."[6]

That process of condensation led Heywood to grapple with questions of classification and values. For example, in considering possible approaches to classifying historical events, Heywood writes, "Of History there be foure species, either taken from place, as Geography; from time, as Chronologie; from Generation as

FIGURE 23. Cover plate of Heywood's *Gunaikeion (1624)*.

Genealogie, or from gests (deeds) really done, which . . . may be called Annologie: The elements of which it consisteth are Person, place, time manner, instrument, matter and thing."[7] Heywood seems to anticipate the kind of multidimensional classification that the great Indian librarian Shiyali Ramamrita Ranganathan would propose centuries later in his famous colon classification (see chapter 9).

Heywood's attempts to liberate his text from fixed top-down categories, as Crook and Rhodes write, "throw into startling relief questions of classification and the logical disposition of knowledge in the period before the establishment of classical taxonomies from the late seventeenth through to the nineteenth centuries." Heywood does more than just call those taxonomies into question. He seems to suggest a prescient alternative to top-down systems of knowledge: a world in which readers actively participate in structuring the universe of available knowledge, rather than following a classification imposed from on high. "The digressiveness of both hypertext and the *Gunaikeion* represents a departure from the kind of narrow, linear thinking that we traditionally characterize as male," write Crook and Rhodes, "and approximates more closely to ways of thinking and learning which in our present culture are considered by some to be the result of women's socialization."[8]

As one of the earliest examples of an encyclopedia-like volume, Heywood's work proved there was a market for such synthetic works. Over the next two hundred years, scholars would find the notion of a meta-book drawn from the contents of other books increasingly alluring. Ultimately, this quest would lead to whole new ways of thinking about the structure of human knowledge and eventually to the great Victorian quest for a universal classification. Those efforts, as we will see in the pages that follow, would eventually find themselves bogged down with increasingly rigid hierarchical constructs; but in their earlier incarnations, encyclopedias seemed to offer the hope of a new, flexible, reader-driven experience. Heywood's notion of a book whose structure is supplied at least in part by its readers seems to foreshadow the self-organizing information ecosystem of the World Wide Web, which, like Heywood's women, seems to resist categorization.

Diderot's Encyclopedia

Well into the eighteenth century, most European households owned at most a single book, often a popular devotional text like the book of hours. Merchants might keep a small private bookshelf with a few volumes, while scholars and clergy might maintain slightly larger private libraries. There was no such thing as a public library. For all the expansion of publishing, most people could not afford many books; they were still expensive commodities. In such a constrained market, the idea of a book culled from other books held out a tantalizing value proposition.

As printed books continued to proliferate across Europe, authors and publishers started to expand their ambitions. By the eighteenth century, a few scholars, inspired in part by Francis Bacon's quest for a unifying philosophical framework, had started to pursue the idea of a great "universal" encyclopedia. History's greatest encyclopedist was Denis Diderot, a Frenchman who adopted Bacon's classification as the foundation for his monumental *Encyclopédie ou dictionnaire raisonné des sciences, des arts et des métiers* (Encyclopedia or dictionary of the sciences, the arts, and the professions), published in a succession of volumes from 1751 to 1772. A massive collection of 72,000 articles written by 160 eminent contributors (including notables like Voltaire, Jean-Jacques Rousseau, and Comte de

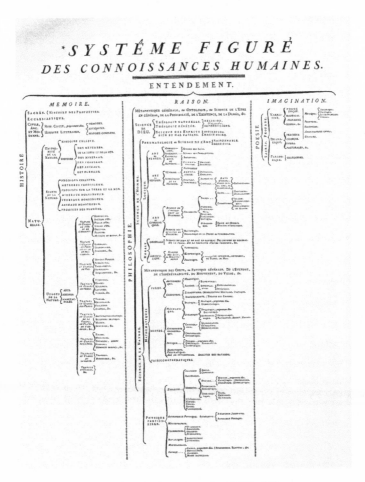

FIGURE 24.   Outline of Diderot's *Encyclopedia* (1751–66).

Buffon), Diderot created a compendium of knowledge unrivaled in any previously published work.

The new encyclopedia was more than just a collection of previously published scholarship. Diderot took the unprecedented step of trying to capture folk knowledge from common tradespeople, devoting an enormous portion of the encyclopedia to knowledge about everyday topics like cloth dying, metalwork, and glassware, with entries accompanied by detailed illustrations explaining the intricacies of the trades. The encyclopedia gave folk knowledge roughly equal billing with the traditional domains of the literary inquiry: scripture, scholarship, and politics. In aristocratic France, such an attempt at mingling high and low cultures marked a provocation. Although publishing this kind of "how-to" information may seem unremarkable today, in eighteenth-century France such an editorial decision constituted a blunt political statement. By granting craft knowledge a status equivalent to the high written culture of statecraft, scholarship, and religion, Diderot issued an epistemological challenge to the aristocratic system. At the time, most lower-class people remained illiterate, and whatever trade knowledge they held still passed primarily through the close-knit social networks of family traditions, apprenticeships, and trade guilds. Almost none of that information had ever been recorded, and even if it had, it certainly would have held little or no interest for the powdered-wig habitués of Parisian literary salons. In according tradespeople such respect, Diderot was sounding a clarion call in favor of the common worker.

Today, we may find it difficult to imagine how the act of publishing information about cloth dying could possibly foment political upheaval. In eighteenth-century France, however, Diderot's work caused enormous consternation among the privileged classes. Pope Clement XIII castigated the work. King George III of England and Louis XV of France also condemned it, anticipating the political consequences of providing the common folk with such easy access to knowledge that had previously been sealed away in tight-knit self-regulated communities. In 1759, the French government ordered Diderot's publisher to cease publication, seizing six thousand volumes, which they deposited (appropriately enough) inside the Bastille.

Diderot died ten years before the revolution of 1789, but his credentials as an Enlightenment encyclopedist would serve his family well in the bloody aftermath. When his son-in-law was imprisoned during the revolution and branded an aristocrat, Diderot's daughter pleaded with the revolutionary committee, citing her father's populist literary pedigree. On learning of the prisoner's connection to the great encyclopedist, the committee immediately set him free.[9]

Diderot's *Encyclopedia* provides an object lesson in the power of printed texts to disrupt old political and religious hierarchies. In synthesizing information that had previously been dispersed in local oral traditions and craft networks, he created a new system that challenged old aristocratic assumptions about the boundaries of scholarship. Just as the networked movement of the French Revolution was powered in large part by the free flow of documents through the revolutionary social network, Diderot drew on the networks of folk expertise to craft a new kind of book that presented an epistemological challenge to the aristocracy.

Today, the encyclopedia is reemerging as a disruptive information technology. In recent years, Wikipedia—a web-based encyclopedia that allows anyone to publish and edit entries—has sparked a fervent debate in scholarly circles. With more than three million entries in more than one hundred languages, it is already far larger than any other encyclopedia ever created. Wikipedia's success has sparked a heated controversy in traditional academic and publishing circles. Questions of authority and credibility inevitably swirl around it. Critics argue that Wikipedia's lack of quality controls leaves it vulnerable to bias and manipulation, while its defenders insist that openness and transparency ensure fairness and ultimately will allow the system to regulate itself. There also seems to be a deeper tension at work, between traditional forms of top-down editorial authority and the bottom-up ethos of the web. Like Diderot's *Encyclopedia*, Wikipedia is stirring tensions between established interests—academic scholars and publishers—and a rising populist sentiment. Although Wikipedia is unlikely to spell the demise of traditional scholarship, it serves as a telling example of the power of "books about books" to pose epistemological challenges to existing institutional systems. The web, like the printing press, seems poised to augur long-term social and political transformations whose effects we are only beginning to anticipate. And once again, the humble encyclopedia may prove the most revolutionary "book" of all.

Aftermath

Three hundred years after Johannes Gutenberg, the printing press had permanently altered Europe's social, political, and intellectual landscape. Ever since Martin Luther posted his theses on the church door, the spread of printing had left a trail of upheaval in its wake. Popular literacy put an enormous strain on old institutional hierarchies. The Vatican had lost its iron religious and political grip, the French government had fallen, and England and Scotland were feeling the first rumblings of the Industrial Revolution. Meanwhile, across the Atlantic, the printing press had provided the essential platform for the world's first fully formed "document nation" (to borrow Brian Stock's phrase[10]): the United States. By the dawn of the industrial era, the Gutenberg revolution had already fostered a profound and lasting transformation of the global information ecosystem.

For scholars, the old monastic ways had fallen into disrepute, and learned readers struggled to develop new secular frameworks to contain the growing sprawl of published knowledge. Despite the best efforts of Renaissance philosophers and encyclopedists, the volume of printed information was outstripping the ability of any one system to keep up, as a burgeoning collection of books and manuscripts threatening to overwhelm all available ontologies. By the end of the eighteenth century, the problem would come to a head in the domain where it all began: taxonomy.

THE MOOSE THAT ROARED

Ay, in the catalogue ye go for men;
As hounds, and greyhounds, mongrels, spaniels, curs,
Shoughs, water-rugs and demi-wolves, are clept
All by the name of dogs: the valued file
Distinguishes the swift, the slow, the' subtle,
The housekeeper, the hunter, every one
According to the gift which bounteous nature
Hath in him closed.

—William Shakespeare, *Macbeth*

In 1787, Thomas Jefferson took delivery in his Paris hotel of the complete skeleton, skin, and antlers of a seven-foot-tall American moose. The moose had journeyed across the Atlantic, packed in salt, and then traveled overland from Le Havre to Paris in a horse-drawn wagon. By the time it arrived at Jefferson's hotel, the animal was in sorry shape. Its skin was sagging, and its hair was falling out. But the thing would serve its purpose well enough. Jefferson, then in Paris as ambassador to France, arranged to have the decomposing quadruped placed on public display in the front entrance to his hotel.[1] Parisians flocked to see the strange spectacle. Jefferson hoped that one particular Parisian, the famous Comte de Buffon, would take note of his specimen.

With this curious stunt, the sage of Monticello cast his vote in a raging scientific debate then captivating Europe's intellectual class. The problem of reconciling the New World with the Old World, as we will see shortly, led to a wave of scientific conflicts that would reverberate for centuries. The moose marked Jefferson's entrée into a taxonomic feud that would have lasting consequences for the future of science, shaping the trajectory of worldwide information systems for centuries to come.

To understand the importance of Jefferson's moose to the subsequent history of information science, we must step back fifty years to early eighteenth-century Sweden, where in 1735 a young botanist named Carl Linnaeus published his landmark *Systema naturae*, a treatise on the classification of plants. Pursuing his lifelong passion for plant life (as a child, he had been nicknamed the "little botanist"), Linnaeus had pursued an ambitious program of collecting information about the plant world and proposing a new systematic framework for describing the natural world, based on his own private studies and on his firsthand experience documenting new plant life while working with the Royal Society of Science. From this project, Linnaeus would emerge as the "father of taxonomy," the first scientist to propose what we would now consider a modern biological classification system.

The eighteenth century was a period of intense research and discovery in the natural world, with scholars discovering thousands of new plants each year. The spread of printing and the rise of the scientific method had fueled a boom in scholarly publishing, as biologists (or natural historians, as they would have called themselves) found the means to share their findings with a growing audience of scholarly peers. Cataloging and sharing their observations, eighteenth-century biologists began to create their own idiosyncratic systems to organize their findings. Absent a unifying conceptual framework, however, much of that work was effectively wasted. Scholars communicated with each other through a vast ad hoc network, publishing books and treatises as they discovered and cataloged new plants in a vast scientific free-for-all. Students of the natural world lived in a Babel of varying naming conventions, assigning Latin words to animal species almost entirely at random. Those names changed frequently as manuscripts changed hands. The hippopotamus—an animal that almost no Europeans had actually seen—went variously by the names river horse, sea horse, behemoth, river paard, and water elephant.[2] And those were just the English names.

The lack of a common classification system made it all but impossible for scientists to build on each other's work. Instead, each scientist labored in a kind of personal echo chamber. Recognizing the problem, philosophers over the years had proposed several attempts at universal classification systems. Joseph Pitton de Tournefot grouped 10,000 species into 689 natural genera. The English naturalist John Ray sorted all known plant species into two groups, monocots and dicots. Some naturalists preferred to use Conrad Gesner's sixteenth-century classification. But none of these systems ever caught on.

By the eighteenth century, publishers were turning out numerous works about the natural world. As naturalists clamored for attention from the growing public readership, the problem of classification was coming to a head. To make matters worse, a stream of fresh data was pouring in from the New World. Travelers were returning from the European colonies with reports of new plants and animals hitherto unheard of: corn, squash, beans, tobacco, tomatoes, cranberries, strange new animals like bison, and variations on existing species like the aforementioned moose. There was even a new "species" of New World human beings to contend with: where did they fit into the tree of life? European naturalists struggled to make sense of the New World.

Linnaeus's *Systema naturae* articulated a simple but powerful system for classifying living things. The system turned on a nested hierarchy, consisting of top-level kingdoms, which in turn were divided into classes, orders, families, genera, and species. For example, within the plant kingdom he identified twenty-four classes, each grouped according to the characteristics of the plants' stamens. Every class was subsequently divided into sixty-five orders, based primarily on the number and orientation of its pistils. Ultimately, his system would encompass more than 7,700 plant and 4,400 animal species.

In the centuries that followed, scientists have expanded Linnaeus's framework to include as many as seventeen levels of hierarchical descriptors, notably adding the phylum (a division between kingdoms and classes) and additional subcatego-

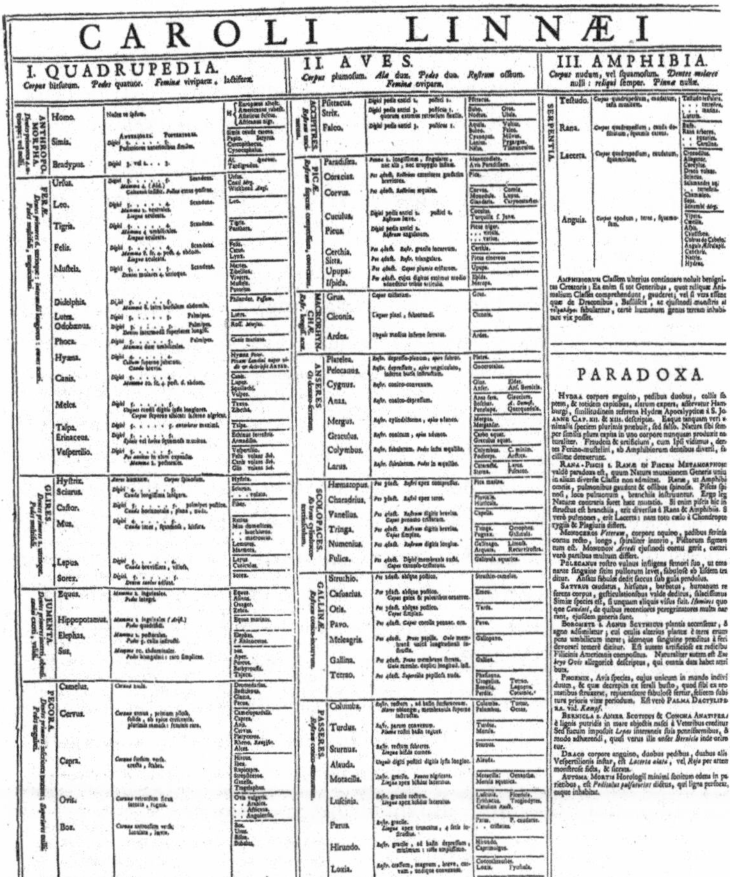

FIGURE 25. Linnaeus's taxonomy (1753). University of Aberdeen.

TABLE 5 Animal Kingdom from Carol Linnaeus's first edition (1735) of *Systema naturae*

LINNAEUS'S CLASSIFICATION	
LEVEL	EXAMPLE
Kingdom	Animals
Class	Mammals
Order	Primates
Family	Hominidae
Genus	Homo
Species	Homo sapiens

ries. But in its original incarnation, the Linnaean system bears a striking resemblance to the folk taxonomies (see chapter 2) that preceded it. This is no accident. Although traditional histories of science tend to give Linnaeus credit as the author of his system, in fact he drew on earlier folk traditions. What sets his system apart was his willingness to embrace folk conventions rather than try to assert his own idiosyncratic vision. Linnaeus intuitively recognized the wisdom of crowds. The subsequent success of his system stems from its roots in our evolutionary heritage. The durability of Linnaean taxonomy may have a great deal to do with the rules governing folk taxonomy: hierarchical categorization of five to six levels deep, binomial naming conventions, and a "real name" or psychologically primary category. The genius of Linnaeus's system owes not just to its transparency but also to its intuitive resonance. For many naturalists, the Linnaean system simply feels "right."

Whereas earlier schemes had stalled at two or three levels of hierarchy, Linnaeus seems to have hit on the precise structure of taxa that occurs among tribal societies. This should come as no surprise; Linnaeus drew heavily on these systems that had persisted through oral transmission since ancient times. Aristotle's naturalist works were already widely known, providing a foundational work for naming conventions (the standard binomial convention of genus species derives from Aristotle), but his great chain of being provided only a one-dimensional hierarchy; it did not provide a unifying theoretical system. Meanwhile, the old folk taxonomies had always persisted in local communities, but although they shared a rough conformity, they were also highly heterogeneous.

Linnaeus's system succeeded by incorporating aspects of folk taxonomies and Aristotelian classification and by articulating an easily understood system that relied not on the mental prowess of individual researchers but on a precise rule set. This rule set in turn lent itself to systematization and to a division of scientific labor. The hierarchical information system enabled an organizational hierarchy to take shape around it. Linnaeus created a team of researchers or, as he called them, apostles, to work under his guidance in fleshing out his classification; each apostle worked to complete a particular piece of the taxonomic puzzle.

By providing a rule set other scientists could follow, Linnaeus paved the way for modern biology. Linnaeus himself allowed that his system was "artificial," using as it did an arbitrary construct of Linnaeus's own devising—the classification of plants according to "male" and "female" characteristics. Linnaeus even used the terms *husband* and *wife*, invoking a matrimonial metaphor that echoes social kinship relations.[3] In fact, Linnaeus hoped someday to supplant his own system with a more "natural" classification that more accurately revealed God's order and spent most of the rest of his career pursuing that elusive goal.

Today, Linnaeus ranks as the uncontested historical founder of modern biological classification. But his ascendance did not come easily. Linnaeus had a powerful rival on the scientific world stage: the French Comte de Buffon, the man who inspired Jefferson to send for his moose.

Buffon's Deist Taxonomy

George Louis LeClerc, the Comte de Buffon was, like Linnaeus, a botanist who had also worked as a royal gardener. And like Linnaeus, he became fascinated with classifying flora and fauna. But unlike Linnaeus, Buffon eschewed strict hierarchies in favor of seeking a "natural" system more reflective of God's will. From 1749 to 1788, Buffon published his monumental *Histoire naturelle*, a compilation of information about known species. The work was a best seller by eighteenth-century standards. Issued in forty-four volumes (eight of which were published posthumously), the work became a standard text in European intellectual circles. Inspired by Pliny the Elder, author of the thirty-seven-volume encyclopedia of the natural world, Buffon rejected earlier Renaissance humanist attempts at reconciling Holy Scripture with observations about the natural world. Instead, Buffon proposed a strictly secular framework predicated on the deist view that God was more of a primal architect who set natural processes in motion, rather than the "hands-on" creator of each individual being.

Buffon believed that humans' propensity for classifying the natural world stemmed from the inherent inadequacy of human perception. In his view, the world existed more as a continuum of life-forms; fixed categories were simply the insecure projections of feeble human minds. Buffon had little use for Linnaeus's detailed classification system. He dismissed the work as an uninspired collection of tables. Buffon's opposition to the Linnaean system of classification, which was based on a narrow set of "essential" anatomical features, would lead him to develop an alternative system. He considered schemes that would take into account an animal's physiology, ecology, functional anatomy (e.g., flying vs. swimming), behavior and geography. This approach ultimately proved unwieldy, but it did make him sensitive to patterns often ignored by Linnaean taxonomists.[4]

Whereas Linnaeus believed in the fixity of species, Buffon was convinced that species were constantly changing; he regarded Linnaean classification as too confining and arbitrary to reflect the true state of the natural world. In this sense, Buffon's approach directly presaged Darwin; indeed, Darwin would later acknowledge his debt to Buffon. But in the intellectual climate of the eighteenth century, Buffon's ideas seemed radically new. Whereas Linnaeus believed that animal species reflected ideal forms that allowed for limited variation that did not change over time, Buffon believed that animals did indeed change in response to their environment, although each animal came preprogrammed with a *moule intérieur* (interior mold) that shaped their development. Since he believed that the natural world was constantly changing, Buffon saw that "artificial" classification systems like Linnaeus's failed to allow for this variation. Buffon found the Linnaean system overly constraining and thought that deriving classification from the characteristics of sexual organisms seemed too one-dimensional, failing to account for the richness and variety of nature. Instead, Buffon wanted to classify animals by their "communities of origin." Buffon was, in other words, an opponent of hierarchical systems. He approached classification from the bottom up, assigning classifications based on empirical observation rather than top-down categories. Most importantly, he believed that these classifications could change over time.

The scholarly rivalry between Linnaeus and Buffon spilled over into personal animosity. Linnaeus chose to name a particularly foul-smelling plant *Buffonia*. Although history has relegated Buffon's theory largely to obscurity, in its day the system commanded an influential following among Europe's learned classes. The struggle between Buffon's and Linnaeus's classifications swirled as one of that age's most active public debates.

Buffon's theory held one important corollary: that because species changed over time, they could also "degenerate" under adverse conditions. The discovery of such degenerate species would mark a resounding affirmation of his theory. Buffon believed he had found just such evidence in the form of the strange specimens coming back from the New World. In *Histoire naturelle*, he wrote, "In America, therefore, animated Nature is weaker, less active, and more circumscribed in the variety of her productions; for we perceive, from the enumeration of the American animals, that the numbers of species is not only fewer, but that, in general, all the animals are much smaller than those of the Old Continent. No American animal can be compared with the elephant, the rhinoceros, the hippopotamus, the dromedary, the camelopard [giraffe], the buffalo, the lion, the tiger, &c."[5]

Further, he noted that the New World appeared to have fewer species altogether—additional proof of that continent's intrinsic inferiority. Given the paucity of species and the apparent deteriorated state of existing specimens, the New World was therefore (his theory went) an intrinsically "degenerate" place.

Which brings us back to Jefferson's moose. As the recently appointed ambassador to France, Jefferson served as booster in chief for the young republic among the Parisian diplomatic set. He squarely faced the challenge Buffon had thrown down. Buffon's doctrine of American degeneracy undercut the young country's international standing. It was, Jefferson wrote, "within one step" of predicting that European immigrants would also degenerate in the New World. If true, what self-respecting European would consider picking up and relocating to such a place?

Jefferson had an especially keen fascination with the natural world. He had personally cataloged all the known plant species in the New World, compiled the 101 species of Virginia birds, filled his home in Monticello with an enormous collection of animal skulls and plant seeds, and kept exacting records of natural events around his home: wind, temperatures, and fifty years' worth of data on the annual springtime appearance of bugs and birds. Indeed, Jefferson was one of the primary contributors to the incoming flow of naturalist reports from the New World. His personal correspondence is peppered with disquisitions on naturalist topics ranging from philosophies of taxonomy to the genitalia of the mole.

In his *Notes on the State of Virginia*, Jefferson went to great lengths to disprove Buffon's contention that American animals were smaller (and therefore inferior) to their European counterparts. Jefferson castigated Buffon's theory of classification as a "no-system." He castigated Buffon as "the great advocate of individualism in opposition to classification. He would carry us back to the days and to the confusion of Aristotle and Pliny, give up the improvements of twenty centuries, and co-operate with the neologists in rendering the science of one generation useless to the next by perpetual changes of its language."[6]

Jefferson embraced the Linnaean classification system, which he felt "offered the three great desiderata: First, of aiding the memory to retain a knowledge of the productions of nature. Secondly, of rallying all to the same names for the same objects, so that they could communicate understandingly on them. And Thirdly, of enabling them, when a subject was first presented, to trace it by its character up to the conventional name by which it was agreed to be called." Although Jefferson did not assert the Linnaean system to be intrinsically more accurate or true than other possible systems, he felt that its legitimacy derived largely from the sheer weight of its popularity.

> I adhere to the Linnean because it is sufficient as a ground-work, admits of supplementary insertions as new productions are discovered,

and mainly because it has got into so general use that it will not be easy to displace it, and still less to find another which shall have the same singular fortune of obtaining the general consent. During the attempt we shall become unintelligible to one another, and science will be really retarded by efforts to advance it made by its most favorite sons. . . . And the higher the character of the authors . . . and the more excellent what they offer, the greater the danger of producing schism.[7]

Jefferson's argument came in the corpse of that moose. It stood at least a third again taller than its continental cousin. This was clearly no "degenerate." It was a tangible refutation of Buffon's theory of American degeneracy and an implicit endorsement of Linnaean classification. Jefferson presented the moose to Buffon personally, along with a copy of his *Notes on the State of Virginia* in which Jefferson demonstrated that domesticated animals were consistently larger in America than in Europe and made a compelling case for biodiversity by demonstrating that America harbored four times more species of quadrupeds than Europe. Buffon was sufficiently impressed with Jefferson's argument that he promised to revise his theories. Unfortunately, Buffon died before completing the task.

Jefferson's endorsement of Linnaean classification would not mark the last time the gentleman from Monticello weighed in on the structure of America's information infrastructure. Jefferson's legacy as a naturalist has long since faded in the shadow of his great contributions as a statesman—as president, vice president, and author of the Declaration of Independence and the Virginia Statute of Religious Freedom. Jefferson, however, prided himself first and foremost on being a naturalist. He is said to have considered it a higher honor to have been named president of the American Philosophical Society than to have been elected vice president of the United States. "Nature intended me for the tranquil pursuits of science," he once wrote, "by rendering them my supreme delight."

Jefferson's Legacy

Almost thirty years after he left Paris, Jefferson would again make a final grand gesture with lasting effects on the structure of information systems. In the fall of 1814 (after he had retired from the presidency), the British sacked Washington, destroying much of the federal government's physical infrastructure, including its nascent library. Just as previous wars had left burning libraries in their wake (see also Alexandria, the Aztecs, and the Chinese emperor Shi Huangdi), the British had tried to burn down the young nation's intellectual capital.

As Washington recovered from the British attack, Jefferson penned a letter to his longtime friend Samuel H. Smith, then serving as chairman of the congressional Library Committee. Jefferson offered his entire personal library as a replacement for the Library of Congress.

Jefferson had amassed an enormous collection, much of it during his tenure in Paris, where he scoured local bookstores and established standing orders with book dealers in Amsterdam, Frankfurt, Madrid, and London. "Such a collection was made as probably can never again be effected," wrote Jefferson. By 1814, Jefferson's library ranked among the largest private book collections in the world: 6,500 volumes, consisting of "what is chiefly valuable in science and literature generally, [it] extends more particularly to whatever belongs to the American statesman. In the diplomatic and parliamentary branches, it is particularly strong." His collection embodied a monumentally important transfer of knowledge between continents.

Fifty years in the making, Jefferson's library at Monticello constituted a formidable intellectual warehouse. Among the volumes Jefferson prized most were John Milton's *Paradise Lost*, Isaac Newton's *Optics*, works by Thomas More and Alexander Pope, numerous histories, and a ten-volume gardener's dictionary. He also had twenty copies of the Bible and two of the Koran.

Congress accepted Jefferson's offer but not without a bout of partisan intellectual wrangling. "The grand library of Mr. Jefferson will undoubtedly be purchased," wrote the *Boston Gazette*, "with all its finery and philosophical nonsense." Massachusetts representative Cyrus King, in high Puritan dunder, insisted that the collection be purged of "all books of an atheistical, irreligious, and immoral tendency." After a bout of acrimonious debate, the gentleman from Massachusetts eventually withdrew his objection. In this small gesture, he may well have influenced the course of history.

Finally, Jefferson loaded his library on ten wagons packed with pine bookcases for the trip from Monticello to Washington. "I hope it will not be without some general effect on the literature of our country," he later wrote.

Even though Jefferson's library was impressive in size, far more interesting was its complex semantic structure. At the time, libraries typically arranged their books by alphabetical order. In describing his own collection, Jefferson explained to the Librarian of Congress George Watterston that he had abandoned the popular convention and chosen as his philosophical reference point "Lord Bacon's table of science," a hierarchical classification scheme that divided the whole of human knowledge into three broad categories: Memory (history), Reason (philosophy), and Imagination (fine arts).

Jefferson modified the scheme to suit his own needs, developing a permuted classification system that would sometimes classify books by subject, sometimes

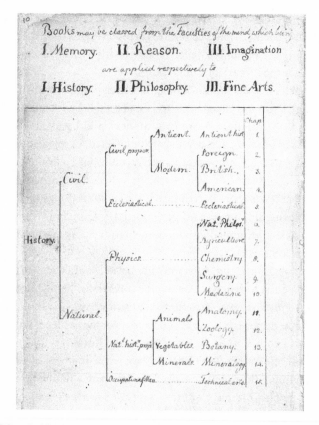

FIGURE 26. Subject matter outline of Thomas Jefferson's personal library catalog (1783). Library of Congress. (See appendix B for the full classification scheme.)

by chronology, "& sometimes a combination of both." It is worth noting that Jefferson's scheme, like so many other hierarchical classification systems, extends five levels deep, echoing the structure of ancient folk taxonomies.

Like Aristotle, Jefferson had created a great personal library, cataloged it, and then donated it as the foundation for a budding national library. And just as Aristotle's own library catalog, coupled with his writings on natural classification, would reverberate long after his passing, Jefferson exerted a similar influence on the intellectual underpinnings of the United States of America. He rarely receives the credit he deserves as a forefather of information science. Jefferson's decisions—embracing Linnaean taxonomy and Baconian empiricism over the competing theories of Buffon and others—set the stage for the information architecture of the young republic.

THE INDUSTRIAL LIBRARY

A library book . . . is not, then, an article of mere consumption but fairly of capital.

—Thomas Jefferson

In 1909, a Boston gentleman named Edmund Lester Pearson remarked on the proliferation of a new apparatus fast becoming a fixture in library reading rooms: the card catalog. "As these cabinets of drawers increase in number," he wrote, "it seems as if the old joke about the catalogues of the Boston Public Library and Harvard University meeting on Harvard Bridge might become literally true."[1]

The image of catalogs colliding seems an apt metaphor for the bibliographical boom that was sweeping the library world. The Industrial Revolution had spawned more than factories, textile mills, and Marxists. It had also transformed the economy of the written word. New steam-driven mechanical presses were stamping out thousands of pages an hour, powering the rise of great publishing houses that turned out books and newspapers in mass quantities, catering to a growing market of newly literate readers (thanks to another industrial innovation: public education). Novels, magazines, pamphlets, and all manner of printed artifacts sluiced through the literary mills. Books became industrial products, even commodities.

Libraries struggled to keep pace. Before the nineteenth century was out, they too would find themselves transformed into industrialized institutions. Pre-industrial libraries had been little more than genteel reading rooms, staffed by learned gentleman curators who looked after modest collections of hand-bound leather volumes. Some university libraries employed a "professor of books," a descendent of the medieval *armarius*, to help guide the bookish student, but there were few "librarians" in the modern sense. Public libraries were all but unheard of; most libraries belonged to universities, learned societies, or private citizens (although some cities boasted private subscription-based libraries for upper-crust society men, like the Boston Athenaeum). Those few libraries with collections large enough to warrant a catalog usually maintained an alphabetical lists of holdings in leather-bound book catalogs. These were typically little more than inventories, designed to keep track of the collection and allow readers to locate a specific book on the shelf.

The industrialization of the printing press and the rise of a growing literate populace exerted severe pressures on these formerly placid institutions. In 1800, the library of the British Museum (precursor to the modern British Library) held forty-eight thousand volumes. By 1833, the collection had quintupled to more than a quarter million.[2] As collections grew, libraries found it increasingly difficult to maintain their old leather-bound catalogs. Faced with rapidly growing collections, libraries inevitably started to mirror the industrial methods that were causing the disruption. Faced with processing backlogs and bulging catalogs, they succumbed to the inexorable logics of industrialization, seeking operating efficiencies, standardizing roles, and normalizing their data. As the nineteenth century progressed, the effects of industrialization would transform the enterprise of the library from a bookman's craft into a bibliographical assembly line.

Today, if you walk into almost any library, you will find what is still, at its core, a nineteenth-century institution. The primary feature of most libraries is still a set of long shelves populated by industrially printed books, organized according to a prescribed hierarchical system of call numbers, maintained by specially trained workers laboring in a highly regimented organizational system. The old card catalog may have given way to computer terminals, but the underlying organizational and ontological structures of the modern library have hardly changed at all. Librarians still follow cataloging practices that originated in the 1850s; library organizations are notoriously top-down, hierarchical, and process-oriented operations. And like steel mills, factories, secondary schools, and other institutional offspring of the nineteenth century, libraries are struggling to reinvent themselves in a postindustrial age.

The conventional history of contemporary library cataloging traces its roots to the nineteenth century, when a proliferation of printed material drove the need for larger, more scalable cataloging systems. But the underlying enabling technology—the catalog card itself—traces its roots back to at least three hundred years earlier, to the mid-sixteenth century. In Switzerland, a naturalist named Conrad Gessner first stumbled on a novel form of note taking. A devoted researcher and obsessive note taker, Gessner would arrange his writing into visual chunks, which he could then cut into small paper rectangles using a pair of scissors.[3] Using this technique, he compiled a massive personal catalog of notes and observations that he relied on to create a series of encyclopedic reference works, the most famous of which—his *Bibliotheca universalis*—attempted to catalog every known book in existence, totaling approximately ten thousand individual books written by some three thousand different writers.

Over the centuries that followed, the technology of card cataloging would gradually evolve as a handful of scholars began to adapt and iterate on Gessner's approach to personal knowledge management. In 1676, Gottfried Wilhelm Leibniz mimicked Gessner's approach with a personal memory storage device that he dubbed the *scrinium literatum* (excerpt cabinet), which he used to document all of his personal notes and observations on small paper slips, which he then organized using a personal classification scheme. Leibniz was fascinated with problems of mathematical abstraction, inventing a series of early mechanical calculators that could add, subtract, and multiple numbers. He even explored hypothetical applications of the binary system (a Chinese framework then largely unknown in Europe), laying the conceptual groundwork for the eventual emergence of microprocessors and modern computing.[4]

The great English encyclopedist James Murray also used a version of Gessner's technique to create the first edition of the *Oxford English Dictionary*. In the 1700s, the French librarian Abbé François Rozier built on Gessner's model to create a rudimentary library card catalog using ordinary playing cards. Gerhard

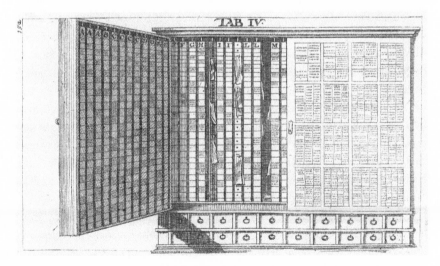

FIGURE 27. Excerpt Cabinet, from Vincent Placcius's *De arte excerpendi*. Reproduced with permission from The Houghton Library, Harvard University. GC6.P6904.689d, tabula IV and V after p. 153.

van Swieten of the Austrian National Library also used an adapted version of the technique to create the so-called Josephinian catalog of the Austrian National Library, consisting of more than two hundred wooden boxes full of bibliographic index cards.

In 1831, the British Museum hired a young Italian expatriate named Anthony Panizzi as its new assistant keeper of printed books. By this time, there were already early rumblings of the industrial printing boom. Books were streaming in and so were readers. In 1759, the library had attracted five visitors a month. A century later, there were 180 visitors a day. Changing conditions seemed to demand a radical rethinking of the library operation. Fortunately for the British Museum, the man they had just hired was no stranger to radical causes. A lawyer by training, Panizzi had fled Italy after escaping arrest for his membership in a secret revolutionary group struggling to liberate Italy from the grips of the Austrian Empire. After a dramatic getaway, Panizzi traveled all over Europe before landing in England. Although he would spend the rest of his life in exile, Panizzi never lost his revolutionary disposition. The fiery young Italian appears to have struck quite a figure among his staid British colleagues. One of his early biographers, Louis Fagan, describes him as "tall, thin and of dark complexion; in temper somewhat hot and hasty, but of calm and even judgment. . . . He must have been most diligent in his pursuit of knowledge, losing no opportunity of study, for he is described as constantly engaged in reading, even while walking from his house to the office."[5] In the years that followed, the energetic Panizzi would stir up more than his share of trouble with the British establishment, bringing his revolutionary zeal to bear in forming a new, highly politicized philosophy of librarianship and leaving behind a legacy of committed, socially engaged librarianship that still resonates today.

By the time Panizzi arrived, the British Museum's library catalog was in desperate need of an overhaul. In 1810, the printed catalog had run to seven volumes; when Panizzi took over in 1831, it had bulged to forty-eight volumes, overflowing with inserts and scribbled errata. When the museum's commissioners asked their young hire to begin work on revising the catalog, they may have thought they were assigning a rote task to a junior employee. But Panizzi, the exiled radical, surprised everyone by coming back to propose the most ambitious revision to the catalog since Callimachus.

Panizzi saw the incoming flurry of books as more than just an inventory problem. He saw the makings of a larger opportunity for the cause of education and literacy. He wanted to reconceive the catalog in a new populist spirit. He began by introducing a subject catalog. Although great libraries of the past had cataloged books by subject, no one had ever tried to create Panizzi's kind of comprehensive classification. He created a new schedule of tiered subject headings, a carefully constructed classification system that reflected the universalist ambitions of the British Empire at the height of its power. "Some would argue [the subject headings] were too ambitious," writes Elaine Svenonius, "that there was no need to construct elaborate Victorian edifices since jerrybuilt systems could meet the needs of most users most of the time."[6] But the introduction of subject headings at all marked a major philosophical step toward opening up the catalog to a broader audience.

In addition to subject headings, Panizzi introduced an important innovation that would allow the catalog to reveal the relationships between particular books. Panizzi explored a sophisticated and nuanced approach to modeling the implicit links between books, recognizing that "a book is a particular edition of a work, a part of a complex web of editions and translations, and that catalog users should be able to see these relationships even as they search for a particular book."[7] He felt the old catalog was too limited, linear, and one-dimensional, proposing instead a new, intricate set of rules for identifying additional "meta"-information such as editions, publishers, dates, and places of publication. He introduced a total of ninety-one rules, covering a vast array of bibliographical scenarios. For example, he introduced the distinction between a "book" and a "work"—that is, between the physical edition and the intellectual property encoded within. That distinction marks a major conceptual leap in the practice of cataloging. By recognizing the dual nature of books as both physical objects and, separately, conveyors of meaning, he introduced an important layer of abstraction. Now librarians could separate the process of identification—locating a particular book on the shelf—and integration—situating the intellectual contents of a book within the topical cosmos of the larger collection. The great Library of Congress cataloger Seymour Lubetzky would later characterize Panizzi's insight as recognizing "the dichotomous character of the book, which, extrinsically, is a separate physical entity, an artifact unrelated to any other, but, intrinsically, a record of human thought and experience."[8] Panizzi's ninety-one rules would become the foundation for the Anglo-American Cataloging Rules, still in widespread use today.

Even though Panizzi's painstakingly detailed rules may seem abstruse and esoteric—the stuff of bureaucratic make-work—Panizzi saw his rules as anything but quotidian. For him, the rules embodied the fulfillment of a higher political purpose. Creating a more usable catalog was no exercise in ontological hairsplitting; it was, in fact, an act of revolution. He wanted to open the library up beyond its traditional audience of privileged men of letters to attract a new class of common readers who initially might know little about books. "I want the poor student to have the same means of indulging his learned curiosity," he wrote, "of following his rational pursuits, of consulting the same authorities, of fathoming the most intricate inquiry as the richest man in the kingdom." He took

FIGURE 28.   The Reading Room and Library at the British Museum, from "London Interiors with Their Costumes and Ceremonies" pub. Joseph Mead, London (ca.1843).

this populist credo a step further, arguing that Britain owed as much to its citizens. "I contend that the government is bound to give the most liberal and unlimited assistance in this respect."[9] Like Denis Diderot, Panizzi was a devout populist willing to stake out a provocative social agenda.

Panizzi's radical vision did not sit entirely well with the British literary establishment. Soon after he introduced his rules, he met with a stout backlash from some of the library's most eminent patrons. One influential reader, Sir Nicholas Harris Nicolas, wrote a withering attack on the new catalog for *The Spectator*, deriding the "exotic capriccios" of the young Italian librarian. Particularly galling to many of the library's longtime readers was Panizzi's insistence on a new procedure for requesting books from the shelves. Previously, readers had simply scribbled book titles on any handy piece of paper and handed it to a desk clerk. Panizzi insisted that readers now had to fill out a printed form, requiring them to consult the catalog to request books by call number (or pressmarks, as they were then known). This trifling bit of bureaucratic process creep sparked a small outrage among some of the library's more prominent patrons. "We know the inconvenience attending this obligation is felt by many literary men," wrote Nicolas, "whose time is of value to them. . . . No more should be expected of a Reader than to specify the book he wants by its title, and that all besides belongs to the librarian." Panizzi seemed to be putting the burden on readers, but in truth he was actually trying to make the structure of the library more transparent so that readers could mine the collection for themselves.

Panizzi's ambitious catalog took far longer than expected to produce, further angering his opponents. Literary worthies like Thomas Carlyle railed against the project: "Elaborate catalogues are not what we require," he wrote. At hearings before the Royal Commission, Carlyle fretted that the library would flounder without a simpler and more quickly produced catalog and that it would become "a Polyphemus without any eye in his head." Carlyle, it should be noted, harbored a personal animus toward Panizzi. He felt that the library staff had failed to support him effectively while he was doing research at the library, and he took his grudge out on Panizzi. Panizzi informed Carlyle that he had received the same level of support as every other user of the library. That came as an outright affront to the famous man, and Carlyle was incensed. He was no ordinary reader after all; he was the most famous writer in England. Despite the ongoing public criticism, the library commissioners stood by Panizzi, eventually forbidding the library trustees from interfering with the library's cataloging decisions and thus ensuring the transformation not just of the British Museum library but of libraries worldwide.

Panizzi's catalog, born of the industrial age, would usher in a new era of democratic access to libraries worldwide. When the catalog was finally published in the 1850s, it became enormously popular as a reference work in its own right. The first edition sold five thousand copies; a subsequent revision was published in twenty thousand copies—a testament to the unprecedented public interest in the new catalog. For all its tribulations, Panizzi's catalog would go down as a landmark in the history of library science. Walk into any public library today and you can still see the visible echoes of Panizzi's revolution at work.

Cutter's Classification

As the industrial wave extended across the Atlantic, American libraries experienced the same growing pains as their British counterparts. Unlike the established great libraries of Europe, America's libraries had no lingering aristocratic class to contend with. From the outset, a populist ethos would guide the development of American library catalogs.

In Boston, the great American cataloger Charles Ammi Cutter pioneered a new way of cataloging books for the public. A devout populist and egalitarian, Cutter believed that the highest purpose of a library catalog should be to serve "common usage," a philosophy that stood in stark contrast to the predominant European view (Panizzi notwithstanding) that library catalogs existed primarily to serve the needs of scholars. Like Panizzi, Cutter viewed his library calling as the fruition of a higher purpose. But Cutter was no political firebrand; he saw cataloging as more of a spiritual pursuit. A former divinity student, Cutter envisioned himself as one of the "humble servants of the Gospel of learning, of knowledge, of science" and espoused his commitment to the "pastoral side of librarianship."[10]

First as a librarian at Harvard and later at the Boston Athenaeum, Cutter introduced a series of innovations. Like Panizzi, Cutter embraced subject cataloging as the key to liberating the contents of the catalog and making them accessible to the common reader. He recognized that unlike scholarly readers, many public library patrons did not arrive with lists of specific book titles; rather, they often came with nothing more than a vague question in mind. The library, then, would succeed or fail based on the usability of its subject catalog.

In his *Rules for a Dictionary Catalog*, Cutter proposed a new multidimensional catalog system, more sophisticated than Panizzi's, which he dubbed the Expansive Classification system. It provided a mechanism for describing books by author, title, and subject, but its main feature was an elaborate multitiered subject scheme. Cutter's system drew distinctions between high-level "classes" (say, history) and "individual subjects" (say, the history of the Peloponnesian War). Using this system, librarians would assign call numbers using letters of the alphabet in one- or two-letter combinations, each letter or combination of letters representing a particular subject; more granular topics could be assigned using sequences of numbers. For example, WP83 stood for United States Painting: "W" for Fine Arts, "WP" for Painting, and "83" for United States.

Cutter's system also recognized the practical limitations of a one-size-fits-all system, acknowledging that different-sized libraries called for different cataloging solutions. He designed his system to scale up or down by releasing it as a seven-tiered system, with the lowest level of classification appropriate for small libraries and the seventh level for large research collections. Despite his hope for a system that would work equally well for small public libraries and large academic collections, Cutter would see his scheme eclipsed by the subsequent success of the Dewey Decimal System, which became the established standard for public libraries. But Cutter's framework lives on in the form of the Library of Congress catalog, also the de facto standard for academic libraries, which uses a two-letter scheme derived directly from Cutter's work. Today, academic library catalogers still assign "Cutter numbers" to new books.

In addition to his breakthrough cataloging work, Cutter introduced a technical innovation that would soon take hold in libraries worldwide: the card catalog. Although he did not invent the card catalog,[11] he recognized its value and advocated it as a logical replacement for the traditional bound catalog. Cutter

TABLE 6   Cutter's Expansive Classification

A	Reference; General works
B–D	Philosophy; Psychology; Religion
E	Biography
F	History
G	Geography
H	Social Sciences
I	Sociology
J	Government, Politics
K	Law
L	General science; Physical sciences
M	Geology; Biology
N	Botany
O	Zoology
Q	Medicine
R	Technology
S	Engineering
T	Manufacturing
U	Military
V	Recreation (sports, theater, music)
W	Fine arts
X	Languages
Y	Literature
Z	Book arts; Bibliography

developed his own makeshift card catalog for the Boston Athenaeum. The device would soon gain a wider following, thanks to the energetic efforts of Cutter's friend and eventual rival Melvil Dewey.

Dewey's Dismal Decimals

Although Cutter likely ranked as the preeminent academic librarian of his day, his historic contribution has been long since overshadowed by Dewey, whose eponymous decimal system is now in use at almost every public library in the United States (and many abroad). Although most nonlibrarians know of Dewey only by virtue of his ubiquitous call numbers, the man's legacy casts a long shadow over the whole enterprise of modern librarianship.

Raised in a small town in western New York by a family of strict Seventh Day Baptist tradition, Dewey's influential life would be marked by what Colin Burke calls "an evangelical drive for reform."[12] While attending Amherst College,

Dewey worked in the college library, where he witnessed firsthand the challenges of a growing library collection and the inefficiencies of the processes then in use for managing the collection. Like most libraries at the time, Amherst organized its books on the shelves into a handful of broad subjects, making it increasingly difficult for library users to locate particular volumes on the shelf. Dewey spent long hours consumed with this problem until he experienced a kind of bibliographical epiphany while attending a Sunday sermon in church.

> For months I dreamed night and day that there must be somewhere a satisfactory solution. After months of study, one Sunday during a long sermon by Pres. Stearns, while I lookt[13] stedfastly at him without hearing a word, my mind absorbed in the vital problem, the solution flasht over me so that I jumpt in my seat and came very near shouting "Eureka!" It was to get absolute simplicity by using the simplest known symbols the Arabic numerals as decimals, with the ordinary significance of nought, to number a classification of all human knowledge in print.[14]

Dewey's supervisors at Amherst became so impressed with his zeal for cataloging that they offered him a full-time job upon graduation, making him the head librarian for the college. Over the years that followed, Dewey single-handedly exerted a profound influence over the trajectory of American libraries. In 1876, he published the first edition of his landmark Dewey Decimal Classification and soon thereafter founded the American Library Association to help support the standardization of practices in libraries around the country. He would go on to found the first library school at Columbia, cofounded (with Cutter) the American Library Association, and set in motion a process of standardization that would enable the great expansion and influence of libraries in American life. Paradoxically, his industrious efforts on behalf of American libraries were predicated in no small part on curtailing the power of librarians.

One of his most visible achievements, the standardized card catalog, serves as a perfect metaphor for Dewey's legacy. Dewey saw the catalog, like the library, as in essence a great machine. By standardizing its operations—introducing interchangeable parts, establishing consistent standards and practices, and normalizing variations—he saw the hope for a more perfect system. Catalog cards, like librarians, would function as distributed cogs in a great national system of Dewey's devising.

Dewey was so obsessed with efficiency that he changed his legal name from Melville to Melvil as a time-saving gesture and briefly even changed his last name to Dui. He once scolded his secretary for wishing him a good morning, admonishing her for such a frivolous use of time that could otherwise have been spent doing work.

Dewey also actively encouraged women to join the emerging library profession, arguing that they were uniquely well suited to the work and that librarianship, as an emerging field, might offer them a faster path to professional advancement that they might find more difficult to obtain in more established male-dominated lines of work. "There is a large field of work for college-bred women in promoting the founding of new libraries," he wrote in an address to the Association of Collegiate Alumnae in 1886. "There is almost nothing in the higher branches which she cannot do quite as well as a man of equal training and experience; and in much of library work woman's quick mind and deft fingers do many things with a neatness and dispatch seldom equaled by her brothers."[15] Although by modern standards, Dewey's views on women would mark him as an unreconstructed chauvinist, his efforts to recruit women into the field played no small part in helping ensure that women had equal access to the profession. By 1930, fully 90 percent of professional librarians were women.[16] To this day, librarianship has taken on a strong cultural identity as a feminized profession, and women still make up the overwhelming majority of American librarians.

Dewey's relentless efforts to create a unified national library system would yield lasting consequences for American libraries. His obsession with efficiency and his strong bent for hierarchical management shaped the trajectory of the entire field of library and information science for more than a century. But his hypercontrolling personality has exerted an unfortunate influence over the subsequent history of American librarians, which have long struggled with excessive bureaucratization and a process-centric work culture that often leaves them struggling to adapt to fast-changing information technologies. Dewey's tireless efforts may have ultimately yielded many public benefits, but it has proved a mixed blessing for librarians, many of whom spend their careers chafing under the stultifying management culture that Dewey played a large part in fostering.

When Dewey introduced his cataloging system for the first time in 1876, he tried to attract interest by touting its economic benefits. "The usefulness of these libraries might be greatly increased without additional expenditure," he wrote, "for with [the Dewey Decimal System's] aid, the catalogues, shelf lists, indexes, and cross-references essential to this increased usefulness, can be made more economically than by any other method which he has been able to find."[17] The appeal to efficiency lay at the heart of his approach. He was devoted less to the arcadian pursuit of human knowledge than to helping libraries manage their operations for maximum productivity and in using them to build a far-flung administrative empire.

For American libraries to move forward, Dewey believed, they had to abandon idiosyncratic local practices and adopt common standards. The economic argument was apparent on its face; the continuing explosion of printed books

demanded efficiency and promised libraries the opportunity to expend their resources on public service rather than maintaining homegrown catalogs. The
adaptation of national standards promised a simple economy of scale. By sup-
planting the unpredictable judgment of willful librarians with the logic of a cen-
trally administered system, libraries could function more efficiently, expanding
their reach by yielding some of their autonomy. It was a classic industrial value
proposition, straight from the factory floor: constricting individual autonomy
in service of the greater ideal of productivity.

In its original incarnation, the Dewey Decimal System called for organizing
books into nine top-level "classes," each with a corresponding beginning number (1 for Philosophy, 2 for Theology, etc.). Each class was further subdivided
into ten "divisions," which in turn were subdivided into one thousand distinct
headings. The system's original divisions included the following:

0 GENERAL
100 PHILOSOPHY
200 THEOLOGY
300 SOCIOLOGY
400 PHILOLOGY
500 NATURAL SCIENCE
600 USEFUL ARTS
700 FINE ARTS
800 LITERATURE
900 HISTORY
(See appendix C for the complete Dewey classification.)

The top-down determinism of the Dewey Decimal System has drawn its share
of critics, who deride the classification as overly simplistic and riddled with cultural bias. Nowhere are its biases more explicit than in its self-evident favoritism
toward Christianity. For example, within the Theology class, divisions 200–280
are devoted to Christian denominations, while Judaism and Islam must make do
with a single number each (296 and 297, respectively). Dewey, in his own defense,
readily admitted that his system was far from perfect. "Theoretically, the division of every subject into just nine heads is absurd," he wrote. "Practically, it is
desirable."[18]

As public libraries started to proliferate across the United States in the late
nineteenth and early twentieth centuries—thanks in no small part to Andrew
Carnegie's generous funding for small-town public libraries—the Dewey Decimal System provided an ideal solution for implementing standardized distributed catalogs across a wide range of library collections. By the middle of the

twentieth century, the Dewey system had cemented its place as the de facto standard for public library catalogs in the United States.

Whatever its shortcomings, the Dewey Decimal System has proved remarkably resilient over the years. Although it undergoes periodic revision, its basic contours have remained unchanged for more than a century. It remains the standard in public libraries throughout the United States (and in many other countries). Although many academic libraries use the more complex Library of Congress Cataloging system, the Dewey system has outlasted numerous competing systems. It has also survived the transition from the old card catalog to the new online catalogs now in service at most public libraries.

Even though physical card catalogs have all but disappeared from most libraries, the imprint of Dewey's industrial library lives on. Most electronic catalogs are in truth simply digitized card catalogs, employing the same bibliographic standards (known in library circles as machine-readable cataloging [MARC] format) and cataloging techniques devised for the old card catalogs. The seemingly modern Internet-connected online catalog is a living relic of the nineteenth century.

In recent years, Internet pundits have invoked Dewey's catalog as an exemplar of the kind of anachronistic, top-down institutional ontology the Internet so effectively subverts. Although Dewey's system surely represents a vestige of industrial age mindset, it is somewhat unfair to dismiss it as simplistically hierarchical. Although the system does provide seemingly rigid categories, it also allows for access through multiple points of entry, such as cross-referenced subject headings; and critics who deride the one-dimensionality of its numbering scheme tend to overlook the fact that the system actually does provide multidimensional access. But the system's enduring strength lies in its simplicity and transparency. Almost anyone can grasp the system; it is easy to implement and easy to use. Like the most successful products of the industrial age, it enjoys mass appeal. As we move into the digital era, however, the value of mass appeal is starting to diminish.

In an era of networked systems, rigid hierarchies are becoming less practical than fluid systems able to adapt to the needs of emergent communities. The industrial virtues of mass appeal and transparency are ceding ground to the digital virtues of flexibility and reconfigurability. On these counts, Dewey's system falls short. In the age of the Internet search engine, the Dewey catalog looks increasingly anachronistic, and almost certainly pressure will continue to mount on libraries to reengineer their old industrial ontologies.

As entrenched as the Dewey system may now seem, it competed with alternative cataloging systems throughout the late nineteenth and early twentieth centuries. Cutter never liked the system, arguing that "its notation would not afford that minuteness of classification which experience taught me to be needed in our library. I did not like (and I do not like) Mr. Dewey's classification."[19] In addition to Cutter's system, other classifications like the Bliss system, the Universal Decimal Classification (see chapter 11), and the Library of Congress system all vied for the attention of libraries; indeed, all of them are still in use somewhere in the world. These systems all shared a similarity of approach, relying on a set of descriptive cataloging rules and a hierarchical schedule of subject headings. They also shared the same fundamental limitation: a reliance on proscriptive subject cataloging. In other words, they relied on the efforts of librarians to identify and anticipate the universe of possible subjects. As the universe of available subjects continued to expand, however, this labor-intensive approach began to show cracks in its conceptual seams.

Ranganathan's Facets

In the 1930s, the visionary Indian librarian Shiyali Ramamrita Ranganathan proposed an entirely new approach to cataloging. Even though his system would fail to catch on, his idea has since proved portentous. As the librarian at Madras University and a prolific writer on library theory (he coined the term *library science* in 1931), Ranganathan wrote a series of books outlining a new model of cataloging known as faceted (or analytico-synthetic) classification. Faceted classification allows for more nuance and specific categories than simple hierarchical subject schemes, by supporting descriptions based on multivalent characteristics, rather than on a single top-down deterministic list of subjects. Ranganathan's colon classification described a framework by which any document could be broken down in terms of five facets: personality, matter, energy, space, and time. These facets could then be combined and interpolated to describe any piece of information in almost endlessly reconfigurable ways.

For example, a faceted classification of a book about management of the London Stock Exchange until 1975 might look like this:

Using the faceted system, subject entries could then be generated by taking any of the major facets as a starting point and interpolating the other facets as qualifiers. It is literally multifaceted. The above example might appear in a catalog in any of the following ways:

Economics.
Stock Exchanges. Management. London. 1975.

London.
Economics. Stock Exchanges. Management. 1975.

TABLE 7

FACET	DESCRIPTION	EXAMPLES
Personality	Primary attribute	Economics
Matter	Physical characteristics	Stock exchanges
Energy	Processes or actions	Management
Space	Location	London
Time	Period or duration	1975

Stock Exchanges
Economics. Management. London. 1975

1975.
Economics. Stock Exchanges. Management. London.

Management.
Economics. Stock Exchanges. London. 1975.

(See appendix E for Ranganathan's colon classification.)

The system thus allows for a multidimensional subject classification that assumes no particular orientation on the part of the user. Using Ranganathan's system, any item can be described in a highly specific way, yet without imposing a deterministic subject hierarchy. It is a bottom-up classification rather than a top-down ontology.

Despite its limited real-world track record, Ranganathan's model has exerted a significant influence, especially on the development of the Library of Congress Subject Headings (LCSH), which support a limited implementation of faceted classification. In recent years, faceted classification has also acquired considerable cachet in computer programming circles, largely because it lends itself so easily to relational database implementations. At the University of California, Berkeley's School of Information, researchers have developed a system called Flamenco that allows users to search and browse a collection of thirty-five thousand images from the fine arts collection cataloged by using hierarchical faceted metadata. Commercial software companies like Endeca are marketing web-based software tools that apply the theory of facets to practical knowledge management applications. Ranganathan's vision has proved particularly well suited to web-based search applications because it lends itself to the kind of iterative querying on which so many web users have learned to rely.

Although Ranganathan enjoys a resurgent reputation among software developers and remains widely admired by librarians, his original colon classification has failed to catch on. The system has never been fully implemented outside

India but remains a kind of platonic object for many librarians and information scientists. Perhaps the simplistic hierarchies of the Dewey and Cutter systems, for all their ontological limitations, exert a deeper psychic pull, invoking our ancient reliance on hierarchical information systems.

Toward a Virtual Library

In 1883, Cutter wrote a futuristic essay titled "The Buffalo Public Library in 1983." Imagining what a library might look like in one hundred years, he envisioned readers sitting at desks equipped with "a little keyboard" through which they could connect with a central electronic catalog, ordering books from the stacks by punching in a call number. He even foresaw networks of libraries connected by a "fonographic foil" that would enable them to communicate telegraphically, accessing each other's collections so readily that "all the libraries in the country . . . are practically one library." He missed the mark by less than a decade.

INFORMATION AS "SCIENCE"

Computer science is no more about computers than astronomy is about telescopes.

—Edsger Dijkstra

One rainy afternoon in 1968, a young Australian graduate student named Boyd Rayward stepped into an abandoned office in the Parc Leopold in Brussels, Belgium. Inside, he discovered "a cluttered, musty, cobwebbed office into which the rain leaked—and one day flooded—causing the attendant then on hand to have a kind of epileptic seizure."[1] Piled high to the ceiling were dusty stacks of books, files, and manuscripts: the intellectual flotsam of a seemingly disorganized old scholar.

The previous occupant, Paul Otlet, had been dead for nearly twenty-five years. A bibliographer, pacifist, and entrepreneur, Otlet in his heyday had been feted as a great man, enjoying the company of Nobel laureates and even playing a role in the formation of the League of Nations. But by the time of his death in 1944, he had lived long enough to see his reputation fade to near obscurity, witness the failure of his greatest ambition, and suffer the final humiliation of watching the Nazis cart away and destroy much of his life's work. When he died a few months before the end of the war, hardly anyone noticed. So ended the legacy of the Internet's forgotten forefather.

In 1934, years before Vannevar Bush dreamed of the Memex and decades before Doug Engelbart and Ted Nelson pioneered the possibility of hypertext information retrieval, Otlet had already delivered a stunningly prescient prophecy of a new kind of networked, multimedia-rich information space:

> Here, the desktop is no longer cluttered. In its place, there is a screen and a telephone port. Off in the distance all the books and information are

stored in a huge edifice, with all the space required for registration and handling with catalogs, bibliographies and indexes, all stored on cards, sheets and folders, with the selection and cataloging handled by a well-qualified staff . . . via wireless, television or telegraphic writing. Data appears on the screen in response to questions sent via telephone or wireless. The screen could split in two, quadruple or increase tenfold if it needed to display multiple texts and documents at the same time; it would have a loudspeaker if the visual material was supported by audio. Wells certainly would have liked this hypothesis. This Utopia does not exist anywhere today, but it could well become the reality of tomorrow if we can continue to perfect our methods.[2]

Otlet imagined a day when users would access databases of recorded information from great distances by means of an "electric telescope," retrieving facsimile images to be projected remotely on a flat screen. At the center of this electronic nexus, Otlet imagined a new kind of scholar's workstation: a moving desk shaped like a wheel, powered by a network of hinged spokes beneath a series of moving surfaces. The machine would let users search, read, and write their way through a vast mechanical database. This new research environment would do more than just let users retrieve documents; it would also let them annotate the relationships between them, "the connections each [document] has with all other [documents], forming from them what might be called the Universal Book."

Otlet's vision rested not just on communications technology but on deep insights into the possibility of stitching otherwise discrete bits of information together, creating semantic relationships that would allow users to navigate from one resource to another across an electronic network. In 1934, the notion of networked documents was still so novel that no one had a word to describe such relationships until Otlet invented one: *links*. He envisioned the whole endeavor as a great *réseau*, a web.

Otlet's idea relied on a new classification framework that, he believed, could succeed where other "universal" classifications had failed. Generations of philosophers had tried to solve the problem of universal classification—including Francis Bacon, Denis Diderot, and John Wilkins—in the nineteenth century. The industrialization of printing, coupled with the advent of cheap binding materials, had spurred an explosion in publishing no less disruptive than the advent of Gutenberg's press four hundred years earlier. Faced with an onslaught of new texts, nineteenth-century scholars and bibliographers had wrestled again with the problem of classification. Catalogers like Anthony Panizzi, Melvil Dewey, and Shiyali Ramamrita Ranganathan had made great headway in systematizing subject headings, laying the foundations of modern library and information sci-

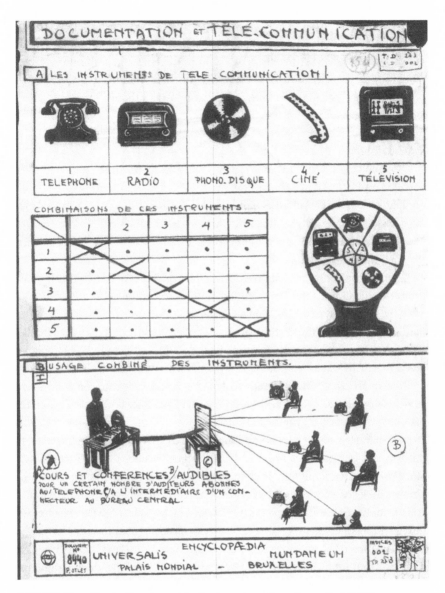

FIGURE 29. Illustration from Otlet's *Encyclopedia universalis* (1936). © Collection Mundaneum, Mons.

ence. With the profusion of printed books, the prospects for a viable universal catalog were growing dim.

Otlet's notion of a networked electronic information network evolved over the course of a long career spent in the pursuit of an idealistic vision that stretched far beyond the scope of information retrieval systems. A committed positivist with a lifelong devotion to the philosophy of Auguste Comte, he believed in the possibility of creating a utopian, post-nationalist state that would transcend the old European order. Ultimately, that vision would entail an ambitious—critics would say impossibly grandiose—plan for a new interconnected world order bound together by a global knowledge network. And although he failed to realize that vision before fading into obscurity in the wake of World War II, his work nonetheless anticipates the networked world that would take shape in the decades to follow.

Otlet took his first step toward trying to unify the world's information in 1895, when he and his partner Henri La Fontaine established the Répertoire bibliographique universel (universal bibliography), an ambitious attempt at developing a master bibliography of the world's accumulated knowledge. Otlet described the project, with characteristic hubris, as "an inventory of all that has been written at all times, in all languages, and on all subjects."[3]

Otlet and La Fontaine recognized from the beginning that the success of the project would depend on the scalability of its classification system. After evaluating the popular systems then in use, including the Dewey Decimal System and Panizzi's scheme for the British Museum, Otlet realized that they all shared a fatal flaw: they were designed to guide readers as far as the individual book but no further. This was an epiphany with potentially revolutionary implications.

Ranganathan had voiced the ethos of modern cataloging when he said, "Every reader his or her book, and every book its reader." But once book and reader were matched, they were left pretty well to their own devices. Otlet wanted to go a step further. He wanted to penetrate the boundaries of the books themselves, to unearth the "substance, sources and conclusions" inside. Taking Dewey's system as his starting point, Otlet began developing what came to be known as the Universal Decimal Classification (UDC), now widely recognized as the first— and one of the only—full implementations of a faceted classification system. Although Ranganathan deserves credit as the philosophical forebear of facets, Otlet was the first to put them to practical use. Today, the UDC comprises over sixty-two thousand classes and has been translated into over thirty languages (one reason for its popularity outside the United States).

In addition to a set of main tables for subject headings, the UDC supports a series of auxiliary tables allowing for the addition of facets. These tables provide notations for place, language, and physical characteristics and for marking relationships between topics using a set of connector signs such as "+," "/," and ":". The UDC's capacity for mapping relationships between ideas—for constructing the "social space" of a document—provides a dimension of use not supported in other purely topical classification schemes. As the Universal Decimal Classification Consortium puts it, "UDC's most innovative and influential feature is its ability to express not just simple subjects but relations between subjects. . . . In UDC, the universe of information (all recorded knowledge) is treated as a coherent system, built of related parts, in contrast to a specialised classification, in which related subjects are treated as subsidiary even though in their own right they may be of major importance."[4] (See appendix D for the current UDC.)

In 1900, in the wake of a successful showing at the Paris World's Fair—where Otlet and La Fontaine won a grand prize for a demonstration of the universal bibliography—the two men began work on an installation at the Palais du cinquantenaire of the Palais mondial. The two men successfully lobbied the Belgian government to provide a home for the operation, which included both the universal bibliography as well as an international association of organizations that Otlet and La Fontaine had cofounded. They secured a wing of a government building in Brussels to incubate the operation, assembling a vast "documentary edifice" of books, newspapers, magazines, photographs, and all manner of other documents—all cataloged on index cards. At the time, the index card represented the latest advance in information storage technology: a standardized, easily manipulated vessel for housing individual nuggets of data. So Otlet's *réseau* began taking shape in the form of an enormous collection of index cards, arranged in a sprawling array of cabinets.

The effort met with early success, even attracting a healthy mail-order business in which users submitted search queries for a fee (27 francs per 1,000 cards retrieved). The service attracted more than 1,500 requests a year on subjects ranging from boomerangs to Bulgarian finance. Over the next few years, Otlet and La Fontaine galvanized support for their efforts, drawing into their orbit a number of prominent thinkers including the Scottish sociologist and polymath Patrick Geddes, the famed modernist sculptor Le Corbusier, and even Andrew Carnegie, who provided funding for a project they dubbed the Peace Palace and expressed support for Otlet and La Fontaine's projects.

Building on the early success of their bibliographic enterprise, Otlet and La Fontaine initiated a series of related, similarly ambitious projects: an international network of associations, a newspaper archive, a network of worldwide museums, and even an entire new city. In partnership with the eccentric Norwegian American sculptor Hendrik Christian Andersen, Otlet and La Fontaine developed a grand plan for the project they dubbed the World City (Mundaneum), which they hoped would serve as the epicenter of a new worldwide government that would usher in a new era of peace, prosperity, and spiritual realization for all of humanity.

As grandiose as the World City project might seem today, it attracted interest from no less a personage than King Albert I of Belgium, who received the three men for a presentation in 1913 and afterward expressed his wholehearted support for the project. "There must be those who dream for the others," said the king. "There is nothing here that cannot be turned into a reality. . . . One day it must exist."[5]

On the eve of World War I, Otlet's fortunes looked promising. His projects had garnered government support and international attention, and his universal bibliography had swelled to more than twelve million entries. When his partner won the Nobel Peace Prize in 1913 for his work supporting the international peace movement, Otlet had good reason to feel optimistic about his prospects. But with war clouds gathering over Europe, his utopian vision of a peaceful interconnected world began to look impossibly optimistic.

Otlet paid a terrible personal price during the war, losing one of his sons in battle and the other to captivity. He shuttered the Palais mondial and fled to Switzerland for a time, where he began to write a series of newspaper editorials urging the formation of a new international body to mediate the world's conflicts. In 1914, he published a long essay titled "The End of War" in which he argued passionately for "a new engine of governance," founded on core principles of human rights and intergovernmental cooperation. For Otlet, such an entity would do more than simply mediate disputes. His proposal called for interconnected legal and political systems, as well as more mundane proposals for things like the metric system, Greenwich Mean Time, calendaring standards, and the adoption of Esperanto as a universal second language.[6] Although Otlet was scarcely the first or only European to imagine a new form of transnational government, he lent a prominent voice to a growing chorus of polemicists whose efforts would lay the groundwork for the eventual formation of the League of Nations in the wake of the war. When the league finally convened its first meeting in 1919, the assembled delegates stood to recognize both Otlet and La Fontaine as two of the leading intellectual lights behind the movement.

Otlet had hoped that the nascent league would choose Brussels as its headquarters (like Switzerland, Belgium had positioned itself as a neutral country in the years leading up to the war)—a development that he hoped would breathe new life into his proposal for a World City. When the league chose Geneva over Brussels as its headquarters, it came as a terrible disappointment—robbing the Mundaneum of one of its primary raisons d'être. By 1924, Otlet's prospects had dimmed further as a more right-leaning government took control in Belgium, showing much less appetite for the grand internationalist projects that Otlet and La Fontaine had been promoting. They had to relinquish the original location, moving the Mundaneum to a succession of smaller quarters, even landing briefly, ignominiously, in a parking garage. After a series of fiscal struggles and management missteps, Otlet finally faced the difficult but unavoidable choice of shutting down the operation in 1934. Six years later, Nazi troops stormed in and carted it all away to make way for an exhibition of Third Reich art.

After Otlet's death in 1944, what survived of the original Mundaneum was left to languish in an old anatomy building of the Free University in the Parc Leopold, all but forgotten for three decades, until Boyd Rayward found his way into Otlet's old office. Over the ensuing half century, more than seventy tons of its original contents were destroyed. Finally, in the mid-1990s, a group of volunteers began resurrecting Otlet's original vision, hoping to preserve and refurbish the original Mundaneum. In 1996, the new Mundaneum opened in Mons, Belgium, to preserve Otlet's legacy and his vision of the "Universal Book." Although today's Mundaneum serves primarily as a museum and learning center rather than as a working incarnation of Otlet's original plan, the new institution does an admirable job of perpetuating his legacy and reminding us of Otlet's premonitory vision of a worldwide networked information environment.

In a bitter irony, the Mundaneum's 1934 closure coincided almost exactly with the publication of Otlet's masterwork, *Traité de documentation*, a manifesto crystallizing forty years' worth of writing and research into the possibilities of networked information structures. Rayward describes the *Traité* as "perhaps the first systematic, modern discussion of general problems of organising information." With the faceted philosophy of the UDC as backdrop, the *Traité* posited a universal "law of organization" declaring that no document could ever be fully understood by itself but that its meaning becomes clarified through mining its relationship with other documents. "All bibliological creation," he said, "no matter how original and how powerful, implies redistribution, combination and new amalgamations."[7]

Even though that sentiment may sound postmodernist in spirit, Otlet was no semiotician; he simply believed that documents could best be understood as three-dimensional things, with the third dimension being their social context: their relationship to place, time, language, as well as other readers, writers, and topics. Otlet believed in the possibility of empirical truth, or what he called facticity—a property that emerged over time and through the ongoing collaboration between readers and writers. In Otlet's world, each user would leave an imprint, a trail, which would then become part of the explicit history of each document.

Vannevar Bush and Ted Nelson would later voice strikingly similar ideas about the notion of associative "trails" between documents. Distinguishing Otlet's vision from the later ideas of Bush, Nelson, and Tim Berners-Lee is the conviction—long since fallen out of favor—in the possibility of a universal subject classification working in concert with the mutable social forces of scholarship. Otlet's vision suggests an intellectual cosmos illuminated by both objective

classification and the direct influence of readers and writers: a system simultaneously ordered and self-organizing and endlessly reconfigurable by the individual reader.

Jorge Luis Borges's fictional Library of Babel was a place containing "all the
possible combinations of the twenty-odd orthographical symbols . . . the translation of every book in all languages, the interpolations of every book in all
books." For Borges, the universal library was a literary conceit. For Otlet, it was
an achievable dream: an "edifice containing all the books and the information
together with all the resources of space needed to record and manage them." Otlet
also recognized the practical importance of "search and retrieval performed by
an appropriately qualified permanent staff." Substitute the word *Google* for *permanent staff*, and Otlet's vision starts sounding a lot like the World Wide Web.

Although it would be an exaggeration to claim that Otlet exerted any direct
influence on the later development of the web, it would be no exaggeration at all to
say that he anticipated many of the problems bedeviling the web today: the explosion of published information, the limitations of fixed documents, and the opportunities for scholarship that open up when readers can penetrate the "meta"-content
surrounding a document to get at the contents inside. In the web's current incarnation, individual "authors" (including both individuals and institutions) exert
direct control over fixed documents. Each document is essentially a fait accompli,
with its own self-determined set of relationships to other documents. It takes a
meta-application like Google or Yahoo! to discover the broader relationships between documents (usually through some combination of syntax, semantics, and
reputation). But those relationships, however sophisticated the algorithm, remain
largely unexposed to the end user, never becoming an explicit part of the document's history. In Otlet's vision, those pathways constituted vital information,
making up the third dimension of social context that made his system so potentially revolutionary.

Would Otlet's web have turned out any differently? We may yet find out. The
explosive growth of social media platforms in recent years suggests that many
web users are clamoring for social context, relying less on formal ontologies than
on socially constructed information spaces. At the same time, many of these social systems suffer from problems of fluidity and instability; they are moving
targets. Otlet's vision holds out a tantalizing possibility: marrying the determinism of a top-down classification system with the bottom-up relativism of emergent social networks.

In Otlet's last book, *Monde*, he articulates a final vision of the great *réseau* that
might as well serve as his last word: "Everything in the universe, and everything
of man, would be registered at a distance as it was produced. In this way a moving

image of the world will be established, a true mirror of his memory. From a distance, everyone will be able to read text, enlarged and limited to the desired subject, projected on an individual screen. In this way, everyone from his armchair will be able to contemplate creation, as a whole or in certain of its parts."[8]

The Documentalists

Although Otlet's work faded into obscurity after the Second World War, the networked world he foresaw would continue taking shape over the decades to come. By the 1930s, an emerging scholarly dialogue had started to take shape in Europe around the transformative social and political possibilities of new forms of documentation. By the mid-twentieth century, these strands of discourse would coalesce into a movement known as documentalism. Whereas library science concerns itself primarily with the archiving and preservation of recorded knowledge, the documentalists saw their remit as shaping the transmission of knowledge throughout society at large. The scientific indexing pioneer Eugene Garfield described the distinction this way: "The documentalist chooses not to be library bound as a modern army tends to be road bound but rather he wants information available to him at all times wherever he may be. Documentation is not only concerned with the organization of materials within the library, but also with making information accessible outside the sacred halls."[9]

The roots of the documentalist movement extend back to the early years of the twentieth century, spurred in no small part by Otlet's expansive utopian universalist vision of worldwide knowledge sharing. In 1915, the German chemist and Nobel Prize winner Wilhelm Ostwalt launched a similarly ambitious project dubbed the Bridge (Die Brücke), an effort to unify the world's scientific information by establishing a new synthetic language for classification. Working in partnership with Adolf Saager and Karl Wilhelm Bührer, he developed a plan for connecting intellectual "islands" like libraries, museums, scholarly societies, and other knowledge-producing institutions. By establishing shared standards for document formats, indexing, and cataloging, the Bridge would provide a central repository for accessing the world's scientific knowledge. He imagined that eventually a scholar might use this system to "unite his field of work with every other field." As more and more scholars used the system, it would evolve into "the great organism of the entire intellectual world."

Ostwalt found a kindred spirit in Otlet. The two men met and corresponded, sharing ideas and ultimately developing a common vocabulary for describing their projects. Ostwalt hoped to connect his Bridge project closely with Otlet's bibliographic enterprise, committing to employing the UDC as its organizing schema.[10] Ostwalt's Bridge project ultimately went the way of the Mundaneum, lost to obscurity in the wake of the mass disruptions that followed World War II.

In the 1930s, a former student of Ostwalt's named Emanual Goldberg developed a system for capturing documents on microfilm and making them easily searchable by means of an indexing system he dubbed the statistical machine, which would enable catalogers to add descriptive metadata using codes entered via a keypad and encoded onto the microfilm, then retrieved using photoelectric cells and digital circuits. Although he secured a patent for the device, he had to halt work on the project after fleeing Germany to escape Nazi persecution in 1938.[11]

Across the English Channel, the famed science fiction writer and social activist H. G. Wells also engaged in the emerging dialogue around the changing shape of knowledge. Like Otlet, Wells saw the flow of recorded information as deeply interconnected with social and political systems and believed that civilization faced a crisis in grappling with the burgeoning phenomenon of information overload. In his 1905 book *A Model Utopia*, he posited the need for reimagining the way people managed information. Comparing society to an organism, he argued that the political tensions facing Europe in the early twentieth century bore a close resemblance to a diseased body with information as its lifeblood. He felt the society could heal itself by optimizing the flow of information. Wells continued to publish and speak on this topic over the next three decades, culminating in his 1939 publication of *World Brain*, a compendium of writing and speeches on the topic of information organization. "Age by age the World's Knowledge Apparatus has grown up. Unpremeditated. Without a plan," he wrote, and "our World Knowledge Apparatus is not up to our necessities." He envisioned that as more and more information began to take shape using new communications technologies, it would eventually form a kind of global encyclopedia, universally accessible to scholars and learning institutions the world over. But realizing that transition would require an enormous technological leap. "Encyclopaedic enterprise has not kept pace with material progress," he wrote. But he saw the promise of emerging technologies like microfilm and electronic devices, tools that he thought might one day offer "a much more fully succinct and accessible assembly of fact and ideas than was ever possible before." Echoing Otlet's prediction of a global network, he predicted that one day "the whole human memory can be, and probably in a short time will be, made accessible to every individual. . . . It can have at once the concentration of a craniate animal and the diffused vitality of an amoeba." Wells went on to describe the worldwide encyclopedia as a kind of "nervous network," one that would ultimately become "woven between all men about the earth."[12] To bring that vision to fruition, Wells envisioned a small army of professional knowledge workers who would comb through the world's published knowledge output, summarizing, cataloging, and unifying related strains of thought into a cohesive whole. Writing in a world long before microchips or artificial intelligence, Wells envisioned a grand synthesis of data happening on a global scale.

Although Wells's utopian vision seems farsighted today, it also contains a darker side. Wells believed that in order for the global encyclopedia to work, it would need a secure and reliable way to identify each of its users. As Boyd Rayward has argued, Wells's vision of a World Brain contains not just an idealistic vision of global knowledge sharing but also a troubling implication that such a system would demand a high degree of social control, "embedded in a structure of thought that may be shown to entail, on the one hand, notions of social repression and control that must give us pause, and on the other, ideas about the nature and organization of knowledge that may well be no longer acceptable."[13]

Wells's musings on the World Brain reveal the totalitarian side of universalist fantasies that seem particularly resonant today: the notion that an increasingly networked world can only function if it tends toward a singular and unified understanding of reality. Wells imagined that the World Brain would ultimately "pull the mind of the world together." More troubling yet, it would beget "a common ideology" that would "compel men to come to terms with one another."[14] As alluring as this fantasy of a unified consciousness might sound, it also implies a totalitarian intellectual regime with little room for dissent or a multiplicity of cultures. In Wells's World Brain, competing perspectives must be subsumed into a singular world order.

Although most of the important work of the documentalist movement was taking shape in Europe, a young science magazine editor named Watson Davis was beginning to take notice of the emerging dialogue in the United States. Originally trained as a physicist, he took on a role as science editor for the *Washington Herald* from 1920 to 1922, before taking on a role as editor of a new syndicated Science Service from Scripps. Trying to stay abreast of a wide range of scientific research gave him a firsthand experience of the problems of information overload in the early twentieth century. And working as a commercial editor also brought him into contact with emerging publishing technologies like microfilm—a development that captured his imagination and led him to become one of the world's foremost proponents of microfilm technology in the 1930s. In 1937, he traveled to Paris to give a presentation on microfilm at the World Congress of Documentation, a landmark event where all of the leading lights of the documentalist movement were in attendance: Otlet, Wells, Goldberg, Ostwald, and hundreds of other librarians and knowledge workers from forty-five countries. The event constituted a high-water mark for the emerging field on the eve of World War II, after which much of the subsequent work on information systems would take on a distinctly military bent in the postwar era. Davis for his part would go on to work closely with Vannevar Bush (see chapter 12) and to play a leading role in the establishment of information science as a field in the United States. He founded the American Documentation Institute, which would later evolve into the American Society for Information Science and Technology, the preeminent professional organization for information scientists in the United States.[15]

Although the influence of the pre–World War II European documentalists on the subsequent development of the postwar computer industry is difficult to pinpoint, Davis provides a tantalizing link to the sequence of events that would ultimately lead to the invention of the Internet and the creation of the World Wide Web.

The Concilium Bibliographicum

Across the Atlantic, a young Quaker student from Pennsylvania named Herbert Field completed his PhD in zoology at Harvard in 1893, at the age of twenty-three. He had already earned a reputation as a brilliant scholar, and his professors hoped he would pursue opportunities as a university professor. But Field struggled with serious physiological challenges including poor eyesight and a severe stuttering disorder. As a result, felt uncomfortable at a classroom podium. Instead, he decided to focus his energies on original research and scholarship, publishing a series of impactful articles that cemented his reputation as an up-and-coming zoologist. As he immersed himself in the burgeoning scientific literature of the time, however, he became more and more concerned with the problem of accessing and archiving published research: a phenomenon he called "the science information problem." Having experienced the challenge of staying abreast of up-to-date research while completing his PhD, he lamented the lack of any reliable indexes or bibliographies of worldwide scientific research. He became so consumed with the problem that in 1893, he announced to his family that he no longer wished to pursue zoology at all but that he was going to devote his energies to the cause of organizing the world's scientific information.[16]

Field's research brought to his attention the work of contemporaries like Dewey, Otlet, and La Fontaine—whom he quickly recognized as fellow travelers. He recognized the efficiency of Dewey's then-revolutionary numeric cataloging system. And he saw enormous potential in the kind of new international bibliographical resources that Otlet and La Fontaine were promoting. Like Otlet, he recognized the potential value of indexing in penetrating deep into the contents of published material to unlock the multiple subjects that might be covered within a specific text. Field also traveled to Russia, where he developed a fascination with the paper-based filing systems in use by the sprawling government bureaucracy. He spent time examining emerging card cataloging systems then in development at the United States Department of Agriculture. And he made contact with a Frenchman named Alphonse Bertillon who was developing a new card-based system for managing criminal records. These encounters helped convince him of both the need and the opportunity for a new centralized index of available scientific information and of the possibilities of card catalogs and numerical classification systems to solve the problem.[17]

Field managed to attract early support from two zoological societies (though he failed to attract funding from larger institutions like the Smithsonian). With some additional funding from his father, he began developing a prototype and a business plan for a new bibliographic enterprise. Taking zoological information as the starting point, he envisioned a subscription-based service in which individual scholars could subscribe to a pack of index cards containing new bibliographic information, to be delivered by mail every two weeks. Ultimately, he envisioned offering highly customized subscriptions depending on individual scholars' areas of interest. He calculated that he would need only five hundred paying subscribers to create a sustainable business operation.[18]

Unfortunately, Field began to encounter a series of obstacles. His health began to fail, slowing progress on the work. And he began to discover the practical challenges involved in incentivizing far-flung scholars to supply voluntary notification of all their recently published work. But he won supporters like the eminent zoologist Julius Victor Carus, and he began to attract paid subscriptions from research libraries around the world. As he began to scale up his operation, he recognized the need to ensure international cooperation and alignment. He made overtures to Otlet and La Fontaine to partner with them more closely and ultimately committed to using Otlet's UDC as the organizing scheme for his index. Unlike Otlet, however, Field did not envision a singular bibliographic institution at the center of the operation; he advocated for a more distributed approach in which subscribers would possess their own collections of catalog cards that would, together, add up to a distributed decentralized system of knowledge.

In hopes of ensuring international acceptance for his project, Field relocated to Paris and gave his enterprise a new Latinate name: Concilium Bibliographicum. He lived alone, working almost continuously on his indexing project in hopes of gaining a toehold in the European research community as quickly as possible. Meanwhile, competitive systems were taking shape—notably from the Royal Society in London, a far better-funded and resource-rich institution, who began work on their own new proposed international classification system. Field recognized that he would need to expand his operation if he hoped to gain widespread institutional acceptance. Soon he began hiring his own professional catalogers to index, classify, and manage the distribution of cards. Thanks to a further infusion of funds following the death of Field's father in 1897, he was able to increase his investment, only to find himself spending 50 percent more than the organization was receiving in income. He nonetheless persevered, investing in a new printing press, professional typesetters, and an automated card-cutting machine. By 1898, the operation was producing and distributing no fewer than 1.6 million cards per year.

Finally, the Concilium reached a point of organizational stability and seemed poised for success. Field had secured long-running subscriptions from major research institutions like the Smithsonian, Harvard, Yale, Princeton, University of Michigan, and the Library of Congress. But for all his evident success, he had failed to account for the skyrocketing volume of scholarship being produced. By the end of 1898, the zoology field alone called for his organization to produce fourteen thousand new cards. But production costs were increasing, and income was failing to keep pace. He struggled to attract enough new subscribers to keep up. Field experimented with new products like more narrowly focused bibliographies or custom "guide" cards providing succinct overviews on particular topics. He also began to branch out to selling cabinets to house catalog cards in Europe (a business model that Dewey had pursued successfully with his Library Bureau in the United States). Taking a page out of Otlet's book, he also decided to create an expensive physical information "palace" in Paris to help telegraph that the Concilium was an institution with organizational staying power. But none of these efforts brought the organization out of the red, and Field continued to deplete his personal fortune.[19]

By 1908, the Concilium began to fall on hard times as subscription growth plateaued and then began to decline in 1909 and 1910. Field considered mortgaging the Concilium building but ultimately knew that he would need more stable sources of funding. He approached Andrew Carnegie in hopes of securing philanthropic support but to no avail. Other fundraising efforts fell flat; finally, Field himself had to take a $50,000 bank loan that he had no means to pay. He grew more desperate and, according to contemporary accounts, mentally unstable. Within a few years, Field would lay off the remaining staff. Finally, Field made the painful decision to leave the Concilium behind, placing it under the care of his secretary.

In the years that followed, Field would go on to an unlikely career of international intrigue. As an early confidant of the American spymaster Allen Dulles, he leveraged his expertise in research methods to help Dulles gather information on emerging conflicts throughout Europe, and he began to establish a strong international network of contacts. Eventually, Field even played a role on behalf of the US government in helping arrange the Treaty of Versailles. After the war ended, Field tried to revive the Concilium project with a renewed focus on fundraising, coming tantalizingly close to gaining institutional support from the Rockefeller

Foundation. Then tragically, Field fell victim to the influenza epidemic then circling the globe, dying of a heart attack just shy of his fifth-third birthday.[20] The Concilium Bibliographicum struggled on under new leadership but would never regain the momentum of its earlier years. The institution shuttered for good in 1939, and with it yet another unrealized dream of a universal bibliography.

Mademoiselle Briet and the Antelope

In 1924, a thirty-year-old librarian named Suzanne Briet, recently graduated from the Sorbonne, reported for her first day on the job at the Bibliothèque nationale, the French national library in Paris. The library's new chief administrator, Pierre-René Roland-Marcel, had started earlier that year and came in determined to modernize the staid institution that for decades had primarily served the country's scholarly elites. He recruited Briet, who graduated first in her class and came highly recommended by her professors. Roland-Marcel was apparently so excited to hire her that he fast-tracked her appointment in spite of pending legislation authorizing him to hire new staff because he saw her as an "exceptional case." Over the next five decades, she would not fail to disappoint.[21]

At the time, the library profession was dominated by men, especially in Europe. Fewer than 10 percent of professional librarians were women (a number that would steadily rise above and beyond 50 percent after World War II). As one of only three women on the library's professional staff, Briet faced considerable headwinds from the more senior male establishment. Soon after she joined, a number of male staff members introduced a motion urging that the library limit the number of female librarians, out of concern that if a woman were to step into a managerial position, she would lack authority over her male direct reports. Fortunately, Briet's new manager, Roland-Marcel, took exception to the motion and made a point of expanding the number of women librarians over the next six years. Over time, Briet and her fellow women librarians won over their peers. As one of their male colleagues E. G. Ledos wrote, the men in the library found themselves gradually won over by their new colleagues' "intelligence, their industry and their conscientiousness." Briet acknowledged these challenges and at one point confided that she wished she had been born male.

Briet entered the field at the cusp of an era of transformational change. In her first year, the Bibliothèque nationale installed electricity, enabling scholars to work past three in the afternoon on cloudy winter days. "It was an unforgettable spectacle," she wrote, "to see the green lamps burst into flower on the tables." That change would pale in comparison to the changes that were coming over the decades to come, many of them driven by Briet herself.[22]

Over the decades that followed, Briet would establish herself as a powerful influence at the library and in the wider professional world, introducing new services and methods that would be widely adopted in other libraries. She introduced a bibliographic instruction service geared toward helping library users navigate the collection, a consultative function bearing some resemblance to the role of reference librarians in the United States but up until that point all but unheard of in European libraries. She argued that library staff had a duty to be "the friend of the reader," motivated by "altruism, team spirit, leadership ability, [and] understanding of the user's psychology." In hopes of propagating this new professional mindset and set of practices, she established a first-of-its-kind training program for documentalists, establishing in 1951 the Institut national des techniques de la documentation, one of the first professional training programs for documentalists that still enrolls hundreds of students every year.

In 1951, Briet published a seminal article titled "What Is Documentation?" that laid out an ambitious view of the field's possibilities and proposed a path forward for an emerging profession. Building on the expansive definitions originally put forward by the first-generation pioneers like Otlet, Wells, and others, she argued that documents are by no means constrained to alphabetic texts but could constitute any form of material evidence: "Is a star a document? Is a pebble rolled by a torrent a document? Is a living animal a document? No. But the photographs and the catalogues of stars, the stones in a museum of mineralogy, and the animals that are cataloged and shown in a zoo, are documents."[23]

By way of example, Briet considers what might happen if a new kind of antelope were captured in Africa and brought back for public display in the Jardin des plantes (botanical garden). The animal itself is placed in a cage, described, and cataloged, and its voice is recorded and archived. From these events, a number of other documentary outcomes would ensue: a press release, resulting in a newspaper article, a radio broadcast, and a mention in a cinematic newsreel; the Academy of Science makes an announcement; a professor mentions the antelope in a lecture; an encyclopedia entry appears; related zoology reference books will need updating; and so on. When the antelope dies, its body would be stuffed and preserved, then put on display in a museum. "The cataloged antelope is an initial document," writes Briet, "and the other documents are secondary or derived." The entire chain of events amounts to a kind of "documentary fertility," and all of these interrelated outcomes are the proper concern of the documentalist.

This expansive definition of documentation marks an ambitious expansion of the scope of the field, suggesting that documentalists should play an active role in shaping the transmission of cultural information across a nearly limitless range of forms and media. Although this definition echoes Otlet's and Wells's early notions of multimedia, Briet goes further and ultimately departs in important ways from the more controlled institutionalist worldview that the early documentalists espoused. She argues for the central role of users in the creation of information services (whereas Otlet and Wells tended to focus more on the needs of institutions and the search for empirical universal truths). And unlike her intellectual predecessors who searched for a singular Aristotelian form of classification that would reveal a unified truth, she takes a decidedly postmodern stance, rejecting the need for "encyclopedic classification schemes" like Otlet's UDC. She criticizes fellow documentalists who "plowed the furrows of a culture that failed, in Otlet's circle" to descend from the clouds," going on to argue that the field had "lost nothing in alleviating itself of a Universal Bibliographic Catalog . . . which everyone had considered a dream."[24]

Instead of a universal cataloging scheme, Briet argued for specialized classification systems purpose built for particular domains but nonetheless coordinated. In calling for this kind of loosely affiliated network of ontological structures, her work anticipates the semantic web and linked data movements that would take shape half a century later. Briet also recognized the possibilities of emerging information technologies to enable humanity to create a new "cultural technique" for organizing and synthesizing information. She saw a kindred spirit in the emerging field of cybernetics at the Massachusetts Institute of Technology, recognizing the "flashy quickness of more effective electrotechnical applications" and foreseeing that the work for the documentalists will become "more and more dependent upon tools whose technicality increases with great rapidity."[25]

The library and information science scholar Ronald E. Day argues that Briet's ideas feel more urgent than ever today in our era of cultural relativism and competing digital narratives. Like Otlet, she had a positivist view of cultural progress and ultimately saw the work of the documentalists as closely aligned with the postwar project of Western international expansion and development. However, unlike Otlet, she resisted the imposition of a singular institutional regime. As Day puts it, "For Briet, Western scientific and Enlightenment values are the seeds through which the world as a whole grows together." Ultimately, she believed transmission of knowledge would start to take shape by, as Day puts it, "networks of multiple documentary-form objects."[26]

Ultimately, Briet's work points the way toward more distributed and localized forms of knowledge production, supported by a technological infrastructure to enable knowledge sharing and collaboration, but one that does not impose ontological uniformity across the world's cultures. For Briet, the library of the future would look less like a hallowed institution and more like a set of widely distributed utility services that enable any number of cultural groups to create their own ways of knowing the world.

Garfield's Proto-Google

As libraries continued to proliferate in the early twentieth century, many of them became more specialized. Driven by a continued explosion of published scholarship and new advances in archiving and indexing tools, a new breed of narrowly focused libraries began to appear: newspaper libraries, business libraries, biology libraries, and other so-called special libraries. Whereas Victorian-era librarians concerned themselves almost exclusively with the organization of books, special librarians saw their roles in a different light, less as book curators and more as active participants in organizational ecologies of information.

In 1915, the American librarian Ethel Johnson wrote, "The main function of the general library is to make books available. The function of the special library is to make information available."[27] In 1909, a group of these librarians had banded together to form the Special Libraries Association. One of its early members, Aksel G. S. Josephson, predicted that eventually librarians "might come to the point where special libraries will not even have whole books, but only such parts of many books as it needs, treating books as well as periodicals on the principle of documentation."[28] That principle, predicated on liberating the contents of books from the confines of their physical volumes—harkening back to Panizzi's original distinction between the book and the work—anticipates the promise of the digital era, the liberation of texts from their physical confines. A new term began to gain currency: *documentation.*

The European documentalist movement galvanized this work and influenced the subsequent development of information science as a discipline in the United States. Across both continents, a new breed of knowledge workers were recasting their traditional responsibilities to take on a more active role in the production of intellectual capital. They no longer saw themselves as cogs in an industrial machine like Dewey's but instead as what we might today call knowledge workers. An early collaborator of Herbert Field named Adolph Voge explored new ways of creating special-purpose indexes and bibliographies, building on Field's classification system. In 1910, the Arthur D. Little librarian Guy E. Marion articulated this emerging vision in his article "The Library as an Adjunct to the Industrial Laboratory" in which he describes the nontraditional "specific nature of the laboratory library" as trafficking in an ever-shifting mix of "textbooks, specialists' pamphlets, trade catalogs, reference works, maps, etc." The mission of the special library extended beyond the acquisition of data from outside the organization, however; these libraries were also deeply involved in collecting data from inside the organization. "There is a vast bulk of data which we may properly call acquired." Internal reports, employee correspondence, laboratory test results, technical reports, and client communications all added up to "the accumulated results of the internal life of the laboratory itself. In fact, this acquired data is unquestionably for the laboratory library the most valuable part of the information."[29]

In the 1950s, a young library science student named Eugene Garfield took up an interest in the problem of managing scientific information, most of which was typically published in specialized journals. Garfield developed a technique known as citation ranking, a method for assessing the impact of scholarly articles by tracking the frequency of bibliographic citations. Depending on the relative frequency of citations, an article would acquire a higher influence ranking, which would in turn lend more weight to whatever documents it cited. The technique, used most powerfully in the influential Science Citation Index, provided a way of ranking the impact of scholarly journal articles. By measuring the cumulative impact of citations, it allowed scholars to assign a collective weight of importance to any given article. The tool effectively circumvented the old manual indexing systems that once predominated in the sciences. The Science Citation Index would go on to exert a sweeping influence on the trajectory of scientific research, becoming the de facto authority in determining the importance of scientific articles and an essential fixture in every science library. At many academic science libraries, a subscription to the Science Citation Index constitutes the single biggest line item in the acquisitions budget.

Inspired by Vannevar Bush's notion of associative trails (see chapter 12), Garfield's notion of citation ranking would in turn inspire two young graduate students at Stanford, Larry Page and Sergey Brin, who extrapolated Garfield's work to the problem of web search results. In their white paper, "The Anatomy of a Large-Scale Hypertextual Search Engine," they outline a new algorithm based on Garfield's idea, Pagerank.[30] That algorithm lives at the heart of their search engine, Google.

Google, for all its success, does not necessarily embody the greatest triumph of Garfield's theory. In fact, Google's approach proves both the power and the limitations of citation analysis. Because it tries to be all things to all people, Google inevitably turns the web into a kind of worldwide popularity contest, often at the expense of smaller, more focused communities of interest. As Paul Kahn puts it, "Link-node models work best on small units."[31] As such, it is particularly well suited for weighing the impact of literature within smaller, focused communities like scientific specialties, where the literature consists mainly of research papers containing references to other papers within a well-defined domain of knowledge. When the same mode of analysis is applied to the great public web, however, problems of context inevitably surface. Without the structured conventions of a scientific paper, the web presents more difficult problems of determining context and the "aboutness" of a particular domain. Google's brute-force algorithm in some ways represents a step backward from the more focused special-purpose research environment that Garfield envisioned.

This emerging role as an information clearinghouse augured a shift of organizational focus for libraries, bringing them directly into the value chain of the industrial enterprise. The emerging special libraries were no longer passive institutions but were now active participants in the production of information. As the twentieth century progressed, innovations like microfilm and punch cards soon opened the door to new ways of working with information.

On the eve of the digital age, the documentalists and their intellectual descendants, the information scientists, had started to anticipate the transformations that would soon reshape the industrial world.

THE WEB THAT WASN'T

> *Come indoors then, and open the books on your library shelves. For you have a library and a good one. A working library, a living library; a library where nothing is chained down and nothing is locked up; a library where the songs of the singers rise naturally from the lives of the livers.*

—Virginia Woolf, *Three Guineas*

So much has been written about the World Wide Web that it scarcely seems productive to try to say anything new about what may be the most documented, and self-documenting, phenomenon in human history. After less than two decades, Tim Berners-Lee's invention already ranks with Johannes Gutenberg's press as a technology of epochal importance. Some futurists have even suggested that the emergence of the web signals an event of evolutionary significance for our species. While it is still much too early to assess its ultimately historical import, there is no question that the web has already left a lasting imprint on the trajectory of human culture.

Conventional histories of computing often have a hagiographic quality about them, celebrating the achievements of notable pioneers (usually white, male, and of European stock) who paved the way forward as part of an inexorable forward march toward progress. These histories often echo a kind of nineteenth-century positivism, undergirded by a belief in the inevitability of progress and the inherent superiority of the Anglo-European technocratic mindset. Even today, the popular press is rife with stories of ostensible technological progress that largely take on this kind of heroic mode of storytelling—the triumphs of Steve Jobs, Elon Musk, Jeff Bezos—or some other tech-savvy protagonist. As we have seen throughout this book, however, information systems always emerge in response to a particular set of social, political, economic, and technological circumstances. The Western archetype of the lone genius can easily blind us to the more complex sociocultural dynamics that shape our information ecosystems. And while the history of the global network is doubtless a story of collective enterprise, nonetheless there were a handful of singular, inventive hypertext pioneers who played a pivotal role in enabling the ongoing information revolution that has come to characterize our age.

The web's astonishing ascent has also lent it an aura of historical inevitability, as if the past century's advances in computing somehow pointed to it all along. If the theory of natural selection teaches us anything, however, it is that evolution rarely runs in a straight line. Just as the fossil record is littered with false starts, dead ends, and unexpected adaptations, so the history of human information systems is largely a story of unpredictable outcomes. Writing emerged as a form of bookkeeping, after all; it took thousands of years for people to start writing down stories. At this early stage of web history, it seems impossible to forecast its ultimate effects. We might do well to remember that throughout human history, the information technologies that mattered most rarely left halcyon outcomes in their wake; more often, they left trails of disruption: burned-out libraries, once-civilized nations regressing into illiteracy, and episodes of blood-curdling violence. For all the popular optimism surrounding today's Internet, a happy ending is far from guaranteed.

Although the web's long-term trajectory may be impossible to project, we can at least look for clues to its near future by mining its recent history. Just as the Gutenberg revolution effectively started well before Gutenberg invented his printing press—thanks to the anonymous efforts of countless unsung medieval monks and bookmakers—so the Internet revolution is really the culmination of broader societal shifts. As far back as the 1890s, a growing flood of published data was already creating the conditions in which a series of new knowledge management strategies would emerge. Those experiments percolated for the next several decades with occasionally promising results but more often leading to frustrating dead ends. As the modern computer industry took shape in the years following World War II, however, computer scientists began developing new hypotheses that broke from the old physical reference points of libraries, catalogs, and indexes to map out a new kind of virtual information space that could take advantage of the emerging digital medium.

Taken together, these early networked technologies not only shed light on how the web came to be, but in many cases, they also present alluring visions of how things might have turned out differently. For all its world-changing importance, the web is still a young technology, one that even now is starting to show a few cracks in the seams. By exploring the history of the "web that was not," we can find glimpses of a web that might have been and, perhaps, clues to a web that may yet be.

Memex Redux

Although Paul Otlet's Mundaneum vanished like a mirage in the wake of World War II, a similarly ambitious vision of automated information retrieval was percolating across the Atlantic. In 1945, Vannevar Bush published his seminal essay "As We May Think" in the *Atlantic Monthly*, sketching the contours of a fictional machine dubbed the Memex.[1] Bush envisioned the device as a scholar's workstation, providing access to large collections of documents stored on microfilm, allowing users to read, write, and annotate documents and, most importantly, to forge "associative trails" between them. Written in the age before digital computers, Bush's essay painted a provocative and inspiring vision of an entirely new kind of machine.

Bush's imaginary device captured the popular imagination right away. Excerpts and feature stories appeared in the Associated Press, *Life*, the *New York Times*, and elsewhere. Even after the initial swell of popular interest subsided, the ideas that Bush suggested would percolate for decades. Today, computer scientists still genuflect reflexively to the inspiration of the Memex. Yet Bush's vision remains surprisingly misunderstood, even by many of those who claim to embrace it. "It is strange that ['As We May Think'] has been taken so to heart in

FIGURE 30. Alfred D. Crimi's rendering of the Memex, in *Life* (1945). Courtesy of Joan D'Amico (Alfred and Mary D. Crimi Estate).

the field of information retrieval," writes Ted Nelson, "since it runs counter to virtually all work being pursued under the name of information retrieval today."[2] Indeed, Bush's essay in some ways reads as an indictment of everything that was about to go wrong with the computer industry.

Although Bush's essay now ranks as one of the canonical texts in computer science—inspiring generations of programmers and, as we shall see later in this chapter, directly influencing the evolution of the World Wide Web—the essay's fame has in some ways overshadowed Bush's real intellectual legacy. "As We May Think" has cast such a long historical shadow that it now obscures much of Bush's subsequent thinking and writing about the Memex, most of which now languishes in out-of-print obscurity. Bush would have no doubt recognized the bitter irony that most of his best writing is nowhere to be found on the web, persisting only on the shelves of the old-fashioned physical libraries he spent much of his career trying to automate out of existence.

The deeper history of the Memex suggests a vision far more provocative than the original microfilm-based device he proposed in 1945. It is not just a story about an innovative idea for managing documents; it is a parable of how a great American engineer turned into cautionary prophet, warning against the influence of corporations on the trajectory of computer science, of the predominance of mathematical and logical models of computing over what he considered a more natural biological approach, and even, in Bush's later writing, espousing provocative and sometimes controversial views on topics like ESP and the co-evolution of human brains and machines. None of these themes would emerge until after the publication of "As We May Think."

To understand the broader implications of the Memex, we need to begin by looking at Bush's earlier experiences in the real world of nonimaginary machines. By the time he started musing about the Memex in the late 1930s, Bush had already designed a series of analog computers dating back to the 1920s, including the successful Differential Analyzer, a room-sized machine that employed a complicated array of gears, cams, and shafts to solve complex mathematical problems like differential equations. While other engineers were probing the possibilities of computing with electrical circuits, Bush took a different, almost nineteenth-century approach, modeling his invention after Charles Babbage's famous gear-driven Difference Engine. Bush's computer did not rely on electrical circuits to perform calculations; it only used electricity to power a set of gear shafts that did the actual math. The thing was a kind of turbocharged abacus.

In 1933, Bush published a long-overlooked essay in *Technology Review* titled "The Inscrutable Thirties." Until this point in his career, he had been a prolific inventor and engineer, publishing a string of dry technical papers and patent filings. Now he began to carve out a more ambitious role for himself as a kind of technological Jeremiah, forecasting the future trajectory of technology by trying to identify long-term predictive trends. Bush scholars James Nyce and Paul Kahn describe the essay as "a literary satire, an ironic description of the present seen from an imaginary future when an appreciation of the 'difficulties surrounding a former generation' would lead the reader to marvel that so much was accomplished with so little."[3] Bush conjures up the character of a harried professor of the 1930s, contending with, among other things, a slow-moving car and a primitive environment for conducting scholarly research:

> The library, to which our professor probably turned, was enormous. Long banks of shelves contained tons of books, and yet it was supposed to be a working library and not a museum. He had to pore over cards, thumb pages, and delve by the hour. It was time-wasting and exasperating indeed. . . . The idea that one might have the contents of a thousand volumes located in a couple of cubic feet in a desk, so that by depressing a few keys one could have a given paper instantly projected before him, was regarded as the wildest sort of fancy.[4]

Bush saw libraries as one of the great unsolved technical challenges of his age, a domain that suffered from unfortunate neglect largely because so much of the nation's technological agenda was driven by corporate interests. Although Bush was by no means opposed to corporations in principle, he recognized that corporations were exerting a growing influence on the development of computing that left less and less room for the kind of human-centered individualistic computing he envisioned. In later years, Bush would lament that the computer revolution had left libraries altogether behind. "The great digital machines of today have had their exciting proliferation because they could vitally aid business, because they could increase profits. The libraries still operate by horse-and-buggy methods, for there is no profit in libraries."[5]

The dream of a digital library was still a long way off in the mid-1930s, when Bush began work on a new kind of computing device, this time designed not for mathematical problem solving but for retrieving bits of information from a storehouse of data. His new Rapid Selector relied on the then-state-of-the-art storage technology of microfilm, allowing users to sift through large collections of documents from a single "terminal" with a projection screen housed in a large desk (unbeknownst to Bush, the German inventor Emanuel Goldberg had designed a strikingly similar photoelectric microfilm selector in 1927).[6] The machine, Bush wrote, "rapidly reviews items on a roll of film, selecting out desired items in accordance with a code. . . . In order to select items from a film, the roll is placed in the machine, a set of indices are placed in accordance with the code of the desired items, and the film is run through rapidly." It was, essentially, a hopped-up microfilm reader with a built-in indexing facility.

By the late 1930s, Bush had built four Rapid Selectors with funding from NCR and Kodak, but they were all plagued with technical problems: they were painfully slow and prone to malfunction. The available technology simply could not keep up with Bush's evolving vision. It was during this period that he began to set his sights further out on the horizon, rather than limiting his vision to the realm of the merely feasible.

In 1939, Bush drafted an essay provisionally titled "Mechanization and the Record" in which he penned almost the entire contents of "As We May Think." The essay sprang in part from Bush's experience as an administrator at the Massachusetts Institute of Technology (MIT), where he witnessed firsthand the growing sprawl of scientific publications threatening to overwhelm the enterprises of science and engineering.

> There is much evidence that we are becoming bogged down today, as specialization extends and research is quickened. There is a growing mountain of research results; the investigator is bombarded with the findings and conclusions of thousands of parallel workers which he cannot find time to grasp as they appear, let alone remember; specialization becomes increasingly necessary for genuine progress, and effort [sic] to bridge between disciplines correspondingly superficial. Still we adhere rather closely, in our professional efforts, to methods of revealing, transmitting, and reviewing results which are generations old, and now inadequate for their purpose.[7]

He went on to describe his proposed solution, revealing an almost complete vision of the Memex fully six years before he published "As We May Think." With war on the horizon, however, Bush shelved his essay to focus on the more imminent threats facing the country. During the war, Bush applied his work on the Rapid Selector to a new machine custom-built for the navy, the Comparator. The war provided him with the impetus to work through the technical kinks that had hamstrung the old Rapid Selector; by the war's end, he had devised a fully functioning information retrieval device. As Bush began winding down his military career, he looked forward to bringing his machine into civilian deployment. "When peace returns," he wrote, "it ought to be applied to something." Turning his eye toward the most challenging information retrieval problems he could find, Bush approached the Federal Bureau of Investigation (whose fingerprint files were sprawling out of control), then the US Patent Office (whose labyrinthine approval and record-keeping processes seemed to make it a ripe candidate for automation), and finally the Library of Congress, whose sprawling collection of books and documents presented an alluring test case.

Once again, the realities of technological constraints and institutional bureaucracies thwarted realization of his plans. The Library of Congress experiment proved at best a mixed success. Partly due to technical limitations (the mechanical film-reading apparatus performed more slowly than expected) but primarily due to the limitations of the manual indexing process, the project sputtered. In the process, Bush developed a crucial insight that would pave the way for the Memex: an antipathy toward manual indexing and toward librarians, who Bush increasingly saw as overly attached to manual processes and a potential obstacle to the user-driven information environment he envisioned.

In 1945, Bush dusted off his 1939 essay and revised it to produce "As We May Think," publishing it first in the *Atlantic Monthly* and later in *Life*. The essay crystallized Bush's first twelve years of thinking about how an ideal information retrieval device would work. The most frequently quoted passage in Bush's essay crystallizes his prophetic vision:

> Wholly new forms of encyclopedias will appear, ready-made with a mesh of associative trails running through them, ready to be dropped into the Memex and there amplified. The lawyer has at his touch the associated opinions and decisions of his whole experience, and of the experience of friends and authorities. The patent attorney has on call the millions of issued patents, with familiar trails to every point of his client's interest. The physician, puzzled by a patient's reactions, strikes the trail established in studying an earlier similar case, and runs rapidly through analogous case histories, with side references to the classics for the pertinent anatomy and histology. The chemist, struggling with the synthesis of an organic compound, has all the chemical literature before him in his laboratory, with trails following the analogies of compounds, and side trails to their physical and chemical behavior. The historian with a vast chronological account of a people, parallels it with a skip trail which stops only on the salient items, and can follow at any time contemporary trails which lead him all over civilization at a particular epoch. There is a new profession of trailblazers, those who find delight in the task of establishing useful trails through the enormous mass of the common record. The inheritance from the master becomes, not only his additions to the world's record, but for his disciplines the entire scaffolding by which they were erected. Thus science may implement the ways in which man produces, stores and consults the record of the race.[8]

The Memex would consist of five interlocking parts: (1) a microfilm document collection; (2) a workstation capable of displaying documents on a projection screen; (3) a mechanism for adding images to the microfilm store; (4) a code input mechanism for identifying and selecting individual records, or groups of records; and (5) associative trails, the conceptual linchpin of the system.[9]

Drawing on his experience with the failures of the Rapid Selector, Bush insisted on the inefficiency of human indexing systems. "Our ineptitude in getting at the record is largely caused by the artificiality of systems of indexing," he wrote. "One has to use rules as to which path will locate [data], and the rules are cumbersome."[10] The basic problem facing contemporary scholarship, as he saw it, was an expanding corpus of literature that "extended far beyond our present ability to make real use of the record."[11] The problem boiled down to, in his view, "the matter of selection," the ability of readers to sift through the available literature, find the most relevant materials, and retrieve them quickly for consultation. He believed that the practice of filing references alphabetically or numerically or by hierarchical classes of subjects forced the reader to operate within a set of cumbersome rules that required enormous cognitive overhead. The journey between documents—central to the scholar's mission—required a constant process of zooming in and out. "Having found one item," he wrote, "one has to emerge from the system and re-enter on a new path."[12]

Like Otlet, Bush sought to liberate the contents of books from the confines of physical volumes, breaking down the old hierarchy of the codex book in favor of a new kind of intertextuality that allowed for direct links between documents, removing the mediating filter of an external index. This kind of internal linking constituted a radical alternative to the deterministic indexing systems then (and still) prevalent in the scholarly publishing world, offering readers a fast path from one document to another. Otlet had called it a link. Bush called it an associative trail. He described it as a model of "selection by association, rather than by indexing." By association he meant allowing authors (and readers) to insert explicit linkages between documents in a collection. "This is the essential feature of the Memex," he wrote. "The process of tying two items together is the important thing." Using associative trails, the user could forge a personal trail through any number of documents, creating an exteriorized representation of an internal thought process that other users could later see. "Thus he goes, building a trail of many items. Occasionally he inserts a comment of his own, either linking it to the main trail or joining it by a side trail to a particular item. . . . He inserts a page of longhand analysis of his own. Thus he builds a trail of his interest through the maze of materials available to him." Like ants leaving pheromone signals for their peers, the scholar could thus create a pathway for others to follow. The associative trail, then, represented a new kind of stigmergy.[13]

Bush's vision differed from Otlet's in two important respects. Otlet's system relied not just on readers' associations but also on the reinforcing framework of a formal classification scheme. Bush eschewed formal classification as needlessly artificial; indeed, he harbored a certain contempt for librarians and social scientists, believing instead that expert scientists could do a better job of self-organizing their own output than a troop of bibliographic bureaucrats. And unlike Otlet, Bush never envisioned the possibilities of penetrating the contents of the documents themselves. Otlet's "links" worked within documents, whereas Bush's "trails" worked only at the higher level of document references, never penetrating the texts themselves.[14]

Whereas Otlet managed to construct a working model of his idea in the Mundaneum, Bush never actually imagined the Memex as a functioning device. "No memex could have been built when that article appeared," Bush later acknowledged. It was from the beginning a conceptual stalking horse, more relevant as a philosophical guidepost than as a blueprint for an actual machine. Indeed, one could speculate that had the machine ever actually been built, its impact might actually have been reduced. With its microfilm reels, photographic copying plate, and forehead-mounted camera, a real-world Memex would today strike most of us as a Rube Goldberg contraption, a historical curiosity rather than the inspirational object it has become.

For all the legions of computer scientists who have since purported to embrace the Memex, few comprehend the depth of Bush's antipathy toward the subsequent trajectory of the computer industry. Even though Bush's Memex has often been associated with the rise of the modern computing industry, Bush never saw his machine as a tool for business. "I proposed a machine for personal use rather than the enormous computers which serve whole companies," he wrote. "I suggested that it serve a man's daily thoughts directly, fitting in with his normal thought processes."[15] Instead, large computer companies have often co-opted his vision in service of big-business agendas.

Paradoxically, even though Bush achieved his greatest results working in large governmental and academic institutions, he was a great believer in individual freedom and saw institutions primarily as the guarantors of environments in which individual genius could thrive. Bush's focus on the individual user ran directly counter to the main trajectory of computing well into the 1970s, when the first personal computers began to emerge. Throughout most of this period, large so-called mainframe computers operated in the back room, executing programs in support of deterministic business or scientific processes. Only a handful of specially trained operators ever had much contact with the machines. In the 1970s, so-called minicomputers gave individual offices and departments their first chance to interact directly with computers. The personal computer revolution of the late 1970s and 1980s brought more people into direct contact with computers, but even the modern personal computer is in many ways a descendant of the old command-and-control militaristic systems. Behind the friendly graphical user interfaces and consumer-friendly marketing messages of the modern PC lies the vestige of exactly the kind of machine Bush did not want to build: a calculating machine, populated by hierarchical information systems like files and folders. Bush's "associative trails" are nowhere to be found in the modern PC. As Nelson puts it, "Bush rejected indexing and discussed instead new forms of interwoven documents."[16]

Even the modern World Wide Web, which does echo certain aspects of his vision, fails to fulfill Bush's larger aspiration for a participatory scholarly research environment. Web browsers are ultimately unidirectional programs, designed primarily to let users consume information from a remote source. To the extent that users can create information through a web browser, they must do so through the mediation of that remote source. As an authoring tool, the browser is all but impotent. The Memex was always envisioned as a two-way authoring environment, allowing users to create new information as readily as they could consume existing information. This is not to say that the Memex is necessarily an entirely superior concept. Bush envisioned the Memex as a powerful standalone machine, while the web works by virtue of a distributed network, lending it the advantages of robustness and scale. And although the Memex may have anticipated the idea of hypertext, it was by no means a true hypertext system. Bush wanted the machine to link individual frames of a microfilm; but given its mechanical limitations, it could not penetrate any deeper into the body of the text to, for example, insert links between individual words or sentences. But Bush's model did enjoy one enormous advantage over the kinds of hyperlinks that prevail across today's web: his links were permanent. By contrast, web hyperlinks are notoriously evanescent.

Although the 1945 model Memex continues to influence developers today, few of Bush's present-day apostles know anything of his later work. In 1945, Bush had spent only a few years thinking about the Memex. "In the quarter-century since then," he later wrote, "the idea has been with me almost constantly, and I have watched new developments in electronics, physics, chemistry, and logic to see how they might help to bring it to reality. That day is not yet here, but has come far closer."[17]

Bush's subsequent vision for the Memex reveals a far more provocative vision, influenced not just by changing technological realities but also by the evolution of Bush's own ideas about how computers should work. In particular, Bush developed a fascination with biological models of computing, as a potential alternative to the mathematical and logical models that have since predominated.

"A great deal of our brain cell activity is closely parallel to the operation of relay circuits," he wrote. "One can explore this parallelism and the possibilities of ultimately making something out of it almost indefinitely."[18] Unlike digital computers, which rely on linear sequencing and the indexing of documents in hierarchical file structures, "the brain does not operate by reducing everything to indices and computation," he wrote. "It follows trails of association, flying almost instantly from item to item, bringing into consciousness only the significant. The trails bifurcate and cross, they become erased in disuse, and emphasized by success. Ultimately, the machine can do this better. Its memory will be far greater and the items will not fade. Progress along trails will be at lightning speed. The machine will learn from its own experience, refine its own trails, explore in unknown territory."[19]

The notion of a Memex that learns, adapting to user behavior along the way, suggests fascinating possibilities of machine-human symbiosis and new ways of thinking about information environments. Today, most of us experience personal computers as fixed entities, with hierarchical folders and a familiar set of programs. Our computers are not so far removed from the dumb terminals of the mainframe era. They know very little about us. Bush's vision suggests the possibility of smarter machines that could anticipate our needs and adapt themselves to our behaviors, like good servants.

Bush also tried to adapt his vision to the changing technological landscape. After the publication of his original essay, Bush recognized that the advent of digital computers made his analog mechanical device seem increasingly anachronistic. Although he never mastered the intricacies of digital computing, he did begin to revise his theories to account for the evolving development of computer technology. In 1958, Bush wrote a sequel to his original essay, titled "Memex II," which was never published during his lifetime.[20] Recognizing that microfilm was a soon-to-be-defunct technology, he envisioned a new kind of memory apparatus using biological crystals. Bush also envisioned the possibility of networking large remote collections of data, of letting users dial in to a Memex from afar. In his final years, he even speculated on the possibility of ESP arising as human brains and external storage devices began to coalesce. He believed that a long-term symbiosis between human brains and computing equipment would lead to the coevolution of new neural pathways in the human brain.

These provocative musings seem all the more prescient today, but Bush's reputation has been largely limited to his comparatively primitive 1945 vision. Although Bush's subsequent thinking remains largely forgotten, "As We May Think" touched off a wave of innovation that still reverberates more than sixty years later.

The Birth of the Information Age

In 1936, Bush hired a twenty-one-year-old master's student at MIT to help him complete work on the Differential Analyzer (see above). That student, Claude Shannon, would go on to publish what historian Herman Goldstine called "one of the most important master's theses ever written, setting in motion a series of events that would prove foundational to the subsequent history of twentieth-century computing."[21]

As a boy growing up in small-town Michigan, Shannon loved tinkering with mechanical devices: experimenting with model planes and boats and even building a rudimentary telegraph system to communicate with a friend across town via a network of barbed wire fences. That mechanical bent would continue throughout his life, as he later devoted countless hours to designing a vast array of machines, including a mechanical mouse, an automaton that played chess (a distant precursor to IBM's Deep Blue supercomputer that defeated chess grandmaster Gary Kasparov in 1997), a trumpet that shot flames, thirty different kinds of unicycles, and a machine for measuring the performance of jugglers called a Jugglometer.

During his undergraduate years at the University of Michigan, Shannon had first encountered George Boole's concept of Boolean algebra in which variables are described not just by letters but also by having qualities of being either true or false. While working on Bush's Differential Analyzer, Shannon applied Boole's concept of true-false encoding to design a series of switching circuits to enable the differential engine to perform complex calculations using a combination of gears and wheels performing precisely choreographed movements.

Building on this work, Shannon built his thesis around the premise that these kinds of Boolean switching circuits could, in theory, be used to solve any number of algebraic equations and drastically simplify the design of electromechanical relays. Shannon's work on the mechanics of binary switching paved the way for the evolution of digital circuit design, the fundamental architectural underpinning of modern digital computers.

In 1940, Shannon moved to Princeton to take a role as a researcher at the Institute for Advanced Study, where he moved in the company of scientific and mathematical luminaries including John von Neumann, Kurt Gödel, and no less a figure than Albert Einstein. Later, he would collaborate briefly with the great British mathematician and World War II code breaker Alan Turing. After the war, Shannon developed an interest in cryptography, leading him to develop the ideas that would ultimately take shape in his landmark work, a journal article titled "A Mathematical Theory of Communication."[22]

Here, Shannon laid the groundwork for contemporary information theory, proposing an approach for encoding information from a sender to a recipient

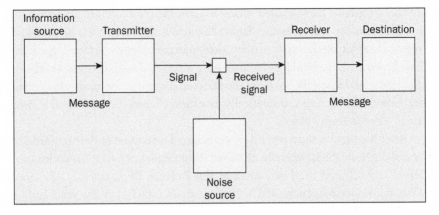

FIGURE 31. Claude Shannon's diagram of a general communication system. Image sourced from Wikipedia Commons.

by converting information into a stream of bits that could be characterized through numerical expression, using a method he dubbed information entropy to assign a statistical representation to describe the amount of information produced by a single letter in a string of text. He also developed a method for encoding these data efficiently for transmission over a circuit.

By 1953, Shannon's work had already started to pierce the public consciousness. In 1953, Francis Bellow wrote in *Fortune* magazine, "It may be no exaggeration to say that man's progress in peace, and security in war, depend more on fruitful applications of information theory than on physical demonstrations, either in bombs or in power plants, that Einstein's famous equation works."[23] If anything, that assertion may have been an understatement. Today, Shannon's work has deeply shaped the digital environment in which many of us engage for many working hours. Nearly all of humanity's current information infrastructure—from mobile phones and laptops to high-speed data networks and satellite systems—all share a common lineage in the coding and data compression methods that Shannon developed.

By 1960, Shannon largely withdrew from publishing and devoted most of his time to tinkering with his beloved devices. But his contributions to the emerging field of data processing and information theory would guarantee his place in computer history.

The Mother of All Demos

Bush's essay would find another enthusiastic disciple in Doug Engelbart, who read the work as a young navy radio technician in the Philippines. "As We May

Think," Engelbart later recalled, would become the philosophical guidepost for his entire career. On returning from the navy, Engelbart set his professional sights on exploring the kinds of networked information retrieval tools that Bush had envisioned. He would go on to develop a series of breakthrough technologies that would shape the course of the personal computer: the mouse, the graphical user interface, and a dramatically new kind of computer system that owed a deep debt to Bush's vision.

After his time in the navy, Engelbart secured a position at the Stanford Research Institute (SRI), where he attracted the attention of J. C. R. Licklider, professor at MIT and then-new head of the Defense Department's Advanced Research Projects Agency (ARPA, later renamed DARPA). A few years earlier, Licklider had penned his landmark book *Man-Computer Symbiosis* as well as a series of memos for what he jokingly described as an "intergalactic network" of connected computers that would eventually unite the nascent academic computer environments at MIT, Carnegie Mellon University, the University of Utah, the RAND Corporation, SRI, and the University of California campuses at Berkeley, Santa Barbara, and Los Angeles. Licklider envisioned a network that would allow geographically distant computers to exchange files with each other, regardless of which types of computers or operating systems they might be using. As Licklider put it, "Consider the situation in which several different centers are netted together, each center being highly individualistic and having its own special language and its own special way of doing things. Is it not desirable, or even necessary for all the centers to agree upon some language or, at least, upon some conventions for asking such questions as 'What language do you speak?'"[24]

Licklider hoped to establish a set of protocols that would allow for worldwide communication between these systems, which he ultimately imagined would pave the way for a global knowledge-sharing system. In Engelbart, he had found a brilliant collaborator who might help him realize this vision.

In 1962, his collaborator Engelbart published a report titled "Augmenting Human Intellect: A Conceptual Framework" in which he described his proposed framework for machine-assisted intelligence. Written during his tenure at SRI for his project sponsors in the air force, the report lays out Engelbart's vision for how an "augmented" human intellect might work. His report was at once a practical blueprint and a philosophical treatise. "By 'augmenting human intellect,'" he wrote, "we mean increasing the capability of a man to approach a complex problem situation, to gain comprehension to suit his particular needs, and to derive solutions to problems." But Engelbart was not referring to the mathematical problems that had preoccupied computer scientists up to that point. He explicitly geared his solution to audiences that had traditionally had no truck with computers: diplomats, executives, social scientists, lawyers, and designers, as well as scientists, like biologists and physicists, who were not then typically using computers. His vision was expansive. "We do not speak of isolated clever tricks that help in particular situations. We refer to a way of life in an integrated domain where hunches, cut-and-try, intangibles, and the human 'feel for a situation' usefully co-exist with powerful concepts, streamlined terminology and notation, sophisticated methods, and high-powered electronic aids."[25] It was, at root, a deeply human-centered vision of computing.

The whole framework rested on a carefully considered model of how humans process information, recognizing that in order to make an information system useful, it must involve breaking down information into atomized nuggets of information that map as closely as possible to the processes of human cognition. "The human mind neither learns nor acts by large leaps," he wrote, "but by steps organized or structured so that each one depends upon previous steps."[26] Engelbart identified the linchpin of the entire system as a series of "process hierarchies" that would enable the system to encode units of information that could be atomized into discrete units of semantic meaning that could then be recombined in endless possible reconfigurations. "Every process of thought or action is made up of sub-processes," he wrote. "Let us consider such examples as making a pencil stroke, writing a letter of the alphabet, or making a plan . . . although every sub-process is a process in its own right, in that it consists of further sub-processes." In turn, each of these process hierarchies forms a building block in a larger repertoire of higher-level hierarchies that Engelbart dubbed a repertoire hierarchy. For example, "the process of writing is utilized as a sub-process within many different processes of a still higher order, such as organizing a committee, changing a policy, and so on."[27] Each of these routines then formed a discrete hierarchical process, which could then be stitched together in any number of possible combinations: networks of hierarchies. The entire framework rests on this conceit.

Working with a team of developers at SRI, Engelbart brought his vision to working reality in the form of a prototype system dubbed the oN-Line System (NLS). Given SRI's institutional priorities and military funding, it is perhaps not surprising that the system placed a special emphasis on facilitating organizational workflow, or what Engelbart called "asynchronous collaboration among teams distributed geographically" (what software developers today would call groupware). The NLS equipped its users with a set of tools that today we would consider rudimentary: a simple word processor, a tool for sending messages to other users, and a rudimentary system for building associated links between documents. The user manipulated a cursor on the screen using a wooden box with wheels attached to the bottom, attached by a wire to the computer. Engelbart's team originally called it a bug but later christened it with a name that stuck: the mouse.

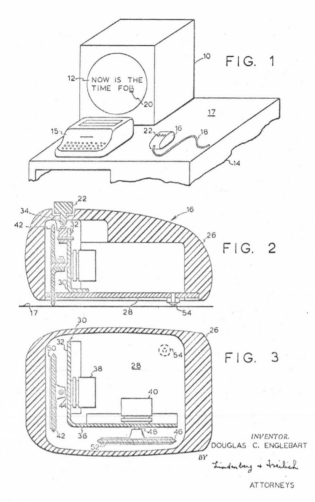

FIGURE 32. Patent Case File No. 3,541,541, X-Y Position Indicator for a Display System, Inventor(s): Douglas C. Engelbart. National Archives and Records Administration.

On December 9, 1968, Engelbart showed a working prototype of his new tool to a public audience at the Brooks Hall Auditorium in San Francisco. In later years, the event would go down in Silicon Valley lore as "the mother of all demos." For many of the one thousand San Francisco audience members that night, his presentation proved nothing short of an epiphany. Perhaps some of them felt something like the curious pagans who first encountered Saint Augustine on the beach wielding his illuminated Gospel.

Dressed in a simple white shirt and tie, Engelbart took a seat in front of a desk bearing a video screen attached to a keyboard and his rudimentary mouse. For the next one hundred minutes, Engelbart proceeded to mesmerize the audience with his demonstration of a system capable of displaying editable words on the screen, exchanging messages with other users, and even a live video conference with one of his colleagues.

Today, many of us may find it difficult to imagine just what a radically different vision Engelbart presented that night. In our age of networked personal computers, cell phones, and digital cameras, it requires a backward cognitive leap to conjure the world of 1968, when computers were still churning out punch cards and tape reels, relegated to back-office tasks managing payrolls, counting election results, and solving complex equations for the National Aeronautics and Space Administration (NASA). Most people had never seen a computer before.

More than a few members of that audience became enthusiastic converts to the digital revolution. Historians of computer science often invoke Engelbart's demo as the seminal event in the personal computing revolution. The NLS found its most immediate successor in a series of projects at the new Xerox Palo Alto Research Center (PARC), a think tank founded with the mission of researching "the architecture of information." Indeed, several of Engelbart's former associates left SRI to join the PARC team. The legacy of PARC on the subsequent development of the personal computer is difficult to overstate. In the 1970s, PARC researchers for all intents and purposes invented today's windows-style graphical desktop, bitmap displays, and even the Ethernet protocol. When a young Steve Jobs visited PARC to see the work in progress, he came away with a vision that would later take shape as the Macintosh (which in turn provided the conceptual foundation for Microsoft Windows).

Engelbart's vision also inspired any number of people outside the mainstream computer industry: people like Stewart Brand, a former Merry Prankster who would go on to found the distinctly hypertext-like Whole Earth Catalog and Whole Earth Review, whose alumni Kevin Kelly and Howard Rheingold would go on to play instrumental roles in the early days of the *Wired* era. Paradoxically, the staid, buttoned-down, military-funded Engelbart found his greatest following in the burgeoning 1960s counterculture. Over the coming years, Engelbart's influence would reverberate throughout the grassroots computer movement that would ultimately unleash the PC revolution.

Slouching toward Xanadu

In the mid-1960s, a young Harvard sociology graduate student named Ted Nelson took his first computer class. He was fascinated by the first-generation computing machines but found the emerging discipline of computer science profoundly dissatisfying. At the time, academic computing was strictly the province of mathematicians, logisticians, and scientists. With a few notable exceptions like Harvard's pioneering Russian-English machine translation project, humanists rarely ventured anywhere near the things. It was a far cry from Bush's vision of an individualized computing experience modeled after the human brain. In the early days of academic computing, mathematicians ran the show.

Soon after he encountered the Harvard computer labs, Nelson started to harbor a gut feeling that the computer scientists had it all wrong. They insisted on approaching computers as glorified calculators. To Nelson, an idealistic young social scientist, a computer "was not a mechanical, a numerical device, it was not a mathematical device, it was not an engineering device, it was an all-purpose machine! . . . Computer was a bad name for it. It might just as well have been called an oogabooga box."[28]

At Harvard, Nelson discovered the germ of an impulse that would propel him into a lifelong pitched battle against the dominant institutional forces of the computer industry. In 1965, he wrote an academic paper titled "A File Structure for the Complex, the Changing and the Indeterminate." In that paper, he coined a word that stuck: *hypertext*. "By 'hypertext,'" Nelson would later write, "I mean non-sequential writing—text that branches and allows choices to the reader, best read at an interactive screen. As popularly conceived, this is a series of text chunks connected by links which offer the reader different pathways."[29] That paper voiced his earliest inklings of a new kind of computing environment designed not for computer scientists but for readers and writers.

Known today as the inventor of hypertext, Nelson has spent most of his career operating outside the industry and academic mainstream. A self-described "rogue intellectual" who self-published most of his work, he nonetheless enjoyed an influential following among a devoted group of early programmers. Computer scientists James Gillies and Robert Cailliau describe Nelson as a Hitchcock-like figure. Just as the great director composed his films while rarely actually looking through the camera lens, Nelson managed to shape the course of the entire computer industry without a formal background in programming.[30] Perhaps it was his love of filmmaking that inspired Nelson to name his pet project Xanadu, just as Orson Welles had named the fictionalized Hearst Castle in *Citizen Kane*.[31]

As a child, Nelson remembers, he reading a magazine called *Flair*, a publication that featured "little doors and windows throughout the magazine you could pop through. You could open the door and you could see something on the next page," he recalls. "I loved that! And so it was obvious that you could get around linearity in these ways but you were always limited physically. The mind could imagine much richer spaces."[32] For Nelson, that childhood template helped guide his thinking about what networked information spaces could be.

Along with Engelbart, Nelson took inspiration from the vision Bush had outlined in "As We May Think." But whereas Engelbart pursued his ideas from within the institutional mainstream, Nelson worked on the fringes, assembling an ad hoc team of programmers to help bring his vision to fruition. Whereas Engelbart tailored his efforts toward teams collaborating on projects, Nelson was more interested in the solitary user. "The fundamental difference between my wonderful and very great stepfather Douglas Engelbart and myself," Nelson says, "is that he wanted to empower working groups and I just wanted to be left alone and given the equipment and basically to empower smart individuals and keep them from being dragged down by group stupidity."[33] In this sense, Nelson is a more direct descendant of Bush, who also advocated for a vision of individual-centered computing.

Nelson was (and is) a devout maverick, intent on shaking up institutional hierarchies with a vision of liberating the individual from suffocating organizational strictures. In Howard Bloom's parlance, Nelson would qualify as a diversity generator, a freewheeling change agent who relentlessly probed the boundaries of the status quo. Whereas Engelbart embraced the notion of hierarchy as the essential building block of his system, Nelson railed against hierarchies of all kinds. "From earliest youth I was suspicious of categories and hierarchies," Nelson wrote; in his books, he railed against the establishment at every turn, dispensing vitriol at two favorite targets: the modern university and the computer industry. "If you are not falsely expecting a permanent system of categories or a permanent stable hierarchy, you realize your information system must then deal with an ever-changing flux of new categories, hierarchies and other arrangements which all have to coexist."[34]

If Engelbart's ideas were a revelation, Nelson's were a revolution. Excoriating the soul-crushing effects of a modern university system that "imbues in everyone the attitude that the world is divided into 'subjects'; that these subjects are well-defined and well-understood; and that there are 'basics,' that is, a hierarchy of understandings."[35] He dismissed professors as "feudal lords" (the duchess of history, the count of mathematics) and went on to vent much of his literary spleen against what he perceived as the professional class of the computer industry. He argued that technologists had created an insular culture for themselves, forming a "polite conspiracy" with humanists based on what he considered a faulty understanding of what computers really were. "Their shared false notion of computers is that they are Inhuman, Oppressive, Cold, Relentless; and that they somehow Reduce Everything to Mathematics."[36] Nelson's aim was to liberate computers from the back office and to transform the whole understanding of what computers were. In this aim, it must be allowed, the computer industry has effectively thwarted his efforts.

Like a latter-day John Ruskin, Nelson railed against the central system ori-entation that then pervaded the computer industry (and, many would argue, still does), with its growing layers of professional specialization. Nelson objected ve-hemently to the emergence of a "priesthood" with a "punch card mentality" that asserted its dominion over the new field with increasing layers of specialist vocabulary, relentlessly excluding nonspecialists, a phenomenon he deemed "the creeping evil of Professionalism, the control of aspects of society by cliques of insiders." Nelson sets out to demystify, and debunk, the institutional computer industry, laying out his new vision of computing centered on the individual, not the organization. Nelson christened himself a systems humanist.

His books, notably *Computer Lib: Dream Machines* and its sequel *Literary Machines*, have in retrospect taken on a visionary glow. But at the time, they went all but ignored among mainstream scholars and the computer industry. In keep-ing with his revolutionary ideals, Nelson wrote in an over-the-top style, em-ploying a mishmash of exotic metaphors, kinetic drawings, and occasionally vituperative diatribes against mainstream institutions that lend his writing a countercultural, occasionally raving, quality. His outlandish style may deter some readers from recognizing the occasional bursts of genius that percolate throughout his work.

The sheer intensity of Nelson's writing presents an imposing barrier to close reading. To wade into a Ted Nelson book is to immerse oneself in turbocharged rhetoric. His language is peppered with wildly inventive words: "zippered lists," "window sandwiches," "indexing vortexes," "part-pounces," "tumblers," "hum-bers," and "thinkertoys." That none of these terms have come close to entering the common parlance testifies to Nelson's secure position on the fringe of the mainstream. But Nelson did invent one word that stuck: *hypertext*.

Computer Lib, released as a reverse-bound, upside-down companion volume to his better-known *Dream Machines*, finds Nelson at his ranting idiosyncratic best. Presaging the *Wired* era by twenty years, *Dream Machines* attracted a cult-like following with its radical alternative vision of computing, outrageous and provocative presentation style, and unapologetic skewering of the institutional model of computing. The book offered readers a technotopian vision of a new world powered by computers but ultimately putting hands in the individual cre-ator. Here, Nelson coined his famous maxim, "Everything is intertwingled," and the book itself seems to reflect that ethos. The text makes every effort to resist linear reading, broken out in a series of cascading chunks and nested diagrams—all delivered in Nelson's signature frenetic voice—evoking the nonlinear quality of "hypertext" that Nelson, at this point, was still only dreaming about. Bound as an oversized manifesto, with three columns of typewritten text interspersed with hundreds of Nelson's hyperkinetic pencil sketches, the whole volume has a handmade mimeographed quality. It almost seems to evoke the nineteenth-century Arts and Crafts ethic, resembling nothing so much as a handmade book. Inside this antic book is a searing vision of what computers could be.

Although today the popular conception of hypertext has been shaped by the working reality of the World Wide Web, Nelson originally articulated a sweeping vision that went far beyond the simplistic model of one-way hyperlinks. Nelson envisioned three basic flavors of hypertext: "ordinary" hypertext, consisting of notational links and references to other documents (roughly the equivalent of a URL in a modern web page); "stretchtext," wherein documents would incorporate entire sections of text by allowing one document to directly access the contents of another and to easily expand one document directly into another; and "collateral hypertext," which facilitated person-to-person sharing of content in which two versions of a document could open on one screen, with a facility for reviewing multiple versions of a document over time. The book also postulates a range of possible semantic relationships going well beyond simple text linking—hypergrams, hypermaps, and "branching movies"—suggesting the possibility of using networked computers to share not just documents but also visceral and sensory experiences. These various permutations of hypertext would also roll up into higher-order collections of knowledge about a particular topic: "fresh hyperbooks," or original works about a particular topic; "anthropological hyperbooks," containing collected references to other hyperbooks, like annotated bibliographies; and "grand systems," consisting of "'everything' written about the subject, or vaguely relevant to it, tied together by editors."

Nelson envisioned all of these threads coming together in a single unified system he dubbed Xanadu, after Samuel Taylor Coleridge's opium-fueled vision of Kublai Khan's imaginary pleasure dome. Imagining his own Xanadu as "a new form of software with potentially revolutionary implications," Nelson envisioned an environment supporting an enormous range of tasks: word processing, file management, email, and the ability to create vertical views of information drawn from a wide repository of sources. Twenty years before the web emerged into the popular consciousness, Nelson outlined a staggeringly ambitious vision:

> A world wide network, intended to generate hundreds of millions of users simultaneously for the corpus of the world's stored writings, graphics and data. . . . The Xanadu system provides a universal data structure to which all other data structures will be mapped . . . a fast linking electronic repository for the storage and publication of text, graphics and other digital information; permitting promiscuous linkage and windowing among all materials; with special features for alternative versions, historical backtrack and arbitrary collaging.[37]

In 1981, Nelson would expand further on his vision of Xanadu in *Literary Machines*, a sequel to *Dream Machines* that would ultimately provide the direct inspiration for the World Wide Web (a debt readily acknowledged by Berners-Lee). By this time, Nelson had spent seven years working with a team of programmers to prototype a real-world implementation of Xanadu. The experience had given him an opportunity to refine his vision and articulate a more detailed one of how the system would actually work: "The Xanadu system, designed to address many forms of text structure, has grown into a design for the universal storage of all interactive media, and, indeed, all data. . . . From this you might get the idea that the Xanadu program is an enormous piece of software. On the contrary: it is one relatively small computer program, set up to run in each storage machine of an ever-growing network."[38]

This vision of a small program (which today we might call a browser) resident on the user's computer interacting with "storage machines" (which today we would call servers) provided the essential blueprint for the operation of the World Wide Web.

Nelson's vision reverberated far beyond the mechanics of client-server networked computers, however. He also proved eerily prescient in anticipating a series of potential concerns with implementing a hypertext environment, such as privacy, copyright, archiving, and version control—all problems very much in evidence in today's implementation of the web. Yet his vision for Xanadu may ultimately have suffered from the complexity of the underlying system, which depended on a system of micropayments and bidirectional links that, if implemented on a global scale, would almost certainly have yielded a complex and unwieldy system that likely would have struggled to gain traction against the simpler, more open architecture of today's web.

Xanadu has nonetheless taken on mythical proportions in the literature of computer science. Despite several attempts, the product has never successfully launched (most recently, the rights had been acquired by Autodesk). But Nelson soldiers on, most recently as a fellow at the Oxford Internet Institute, where he continues to pursue his inspired and idiosyncratic vision of an alternative universe of humanist computing. Despite his outsize influence on the evolution of the web, Nelson remains surprisingly little-read today. *Dream Machines* has long been out of print; copies now fetch $300 or more (I was fortunate enough to find a dog-eared copy kept under lock and key at the San Francisco Public Library). Although Nelson has worked with large corporations and institutions in trying to bring his vision to life, his colorful style and virulent anti-establishmentarianism seem to have guaranteed his status as a committed outsider.

In the struggle between networks and hierarchies, Nelson has cast his lot squarely on the side of the network, insisting on a purely bottom-up model of information storage and retrieval, absent familiar hierarchies like files and fold-ers, and liberated from the control of the computer priesthood that he has de-voted his career to confronting.

Nelson's hopes for a humanist vision of computing may ultimately amount to so much tilting at windmills. For all its promise of individual creativity and liberation, the control of today's web ultimately rests in the hands of corporate and governmental entities. Although many scholars have decried the commer-cialization of the web, however, perhaps we can take some solace in knowing that every major information technology in human history has taken hold pri-marily due to commercial impetus. Writing emerged at the hands of merchants; the printing press spread not because of the Gutenberg Bible but on the strength of a booming business in religious indulgences and contracts. So it should per-haps come as no surprise that adoption of the web has been fueled largely by commercial interests. But Nelson's work still resonates as an inspired, if at times hyperventilated, alternative vision of a humanist computer environment.

Beyond Xanadu

While Nelson and his band of idealistic programmers were pursuing their vi-sion outside the institutional mainstream, the burgeoning computer industry was growing largely at the behest of large organizations like banks, insurance companies, the government, and the military. Computers lived in the so-called back office, tended by specialized information technology staff—Nelson's "priesthood"—who typically had little direct contact with the front office mar-keting, sales, and management staff. Corporate computing followed a strictly hierarchical model in which systems architects would carefully mediate the re-quirements of the front office into structured documents that formed the basis for rigid, deterministic programming regimens designed to maximize efficiency and minimize risk. Information flowed in tightly controlled channels. The age of the desktop PC was still over the horizon.

The command-and-control model of enterprise computing fueled the indus-try's growth in the postwar era, but during the late 1960s and 1970s an alternative vision of human-centered computers began to percolate. In Palo Alto, California, a group of former Engelbart disciples came together at the legendary Xerox PARC, a research and development facility that Xerox founded in 1970 with the express mission of creating "the architecture of information." That group's accomplish-ments, including the graphical user interface, What You See Is What You Get (WYSIWIG) text editing tools, the Ethernet protocol, the laser printer, and the pioneering Alto personal computer, have been amply chronicled elsewhere.

By the early 1980s, many universities began to experience the emergence of two separate computing cultures: the administrative information technology infrastructure of the formal organization that handled the payroll, ran the online library catalog, and processed billing, payments, and course registrations. These were Nelson's "priests." Typically working behind an organizational firewall, the information technology department usually maintained its own battery of mainframes and minicomputers. End users, in this culture, still used green-screen "dumb" terminals, but all the actual computing happened behind closed doors. Meanwhile, in the computer labs, faculty offices, and dorm rooms, students and faculty were starting to build their own computing infrastructure, often operating outside the administrative information technology department. Whereas in the 1970s and 1980s everyone had shared space on the campus mainframe, now the two cultures were beginning to diverge: minicomputers and mainframes for the administration and Unix workstations and personal computers for the academics. Students and faculty started using PCs and sharing information with each other.

Across the continent, another group of researchers explored the possibilities of networked information retrieval in an academic setting. In the late 1960s, Andries van Dam led a research team at Brown University in producing the first working prototype of an interactive hypertext system on a commercially available mainframe, the IBM 360/50. Inspired by a chance meeting between Van Dam and his Swarthmore contemporary Ted Nelson at the 1967 Spring Joint Computer Conference, the system came to be called the Hypertext Editing System (HES). Nelson helped Van Dam formulate requirements for the system, designing the hypertext features and contributing a set of interconnected patents on electroplating. Eventually, Nelson became disappointed with the project's emphasis on paper printout capabilities (by Nelson's own account, the Brown researchers viewed him as a "raving" and "flaming" intruder), and he left Brown to pursue his hypertext vision in the form of his new Xanadu project. Nonetheless, HES marked an important early validation of Nelson's vision: it proved that a hypertext environment could work in practice. Meanwhile, Van Dam went on to build a series of pioneering hypertext systems that would ultimately be used by thousands of people around the world.[39]

The original HES system had two equally important goals: (1) to support individual authors in the process of creating documents or other work products and (2) to explore the notions of nonlinear information structures that Nelson had dubbed hypertext. These goals arose out of the needs of a user community that wrote papers on typewriters and had no interest in working with the teletype-driven text editors that programmers used to write programs at the time. Nelson's vision of hypertext suggested a new way of working with ideas that would be free of the linear constraints imposed by the then-ubiquitous paper media.

HES ran on standard hardware and software (IBM 360/50 OS/MFT with a 2250 graphics display terminal), combining nonhierarchical and flexible document structures with the first true non-line-oriented word processing system. The Brown team envisioned HES as a research system, intended to explore notions of document structures and text editing; as such, it did not immediately come into wide use on campus, especially since it was quickly replaced by research on its successor, the File Retrieval and Editing System (FRESS), a system designed to overcome the limitations of HES, extend its strengths, and incorporate lessons learned from Engelbart's NLS. Nonetheless, IBM released it as part of its Type-III Library of customer-generated applications. Years later, Van Dam would receive a letter from a NASA program manager, informing him that HES had been used to produce Apollo documentation that went to the moon.

In 1967, almost as soon as HES was deployed, van Dam and his team envisioned a more accessible tool, one that could be employed across campus. The result was the File Retrieving and Editing System (FRESS). Once again, the team had designed a computing environment as both a process and a product. When HES development started, Van Dam had not known about Engelbart's work with the NLS, but by the time FRESS development was under way, he did. Consequently, FRESS was a response to both the experience with HES and the insights gained from the NLS, both positive and negative.

FRESS was a multiuser device-independent system. It ran on equipment ranging from typewriter terminals and glass teletypes to high-end vector graphics terminals with light pens. The system was designed to support online collaborative work, the writing process, and document production. It had bidirectional links, macros, metadata in the form of both display and textual keywords, and a notion of document structures that in some sense paralleled programming structures, although it lacked fully conditional branching, multiple windows, view specifications, structured links that allowed documents and document views to be dynamically structured on the fly, and dynamic access control mechanisms that accommodated conflicting demands on the same information corpus.

Users could include two kinds of links: "tags" pointing to another part of the same document, such as an annotation or footnote, and "jumps" that provided bidirectional links between documents. Developed in the years before the computer mouse became a common feature, the system at first allowed users to manipulate objects using a light pen and a foot pedal ("point and kick," to borrow Gillies and Cailliau's pun).

As primitive as the system might have seemed with its foot pedals, green screen, and text-only readouts, FRESS was in some ways more sophisticated than today's web. Links worked in both directions—a link from one document to another would always invoke a reciprocal link going the other way. If someone linked to your document, you would know about it. For example, a class of students could annotate a set of documents, adding their own links, and view whichever links they wanted to see from their peers. FRESS also provided a facility for keyword annotations, allowing readers to search and filter by subjects and to assign names to discrete blocks of text for later research. The ability to assign free-form keywords—now often referred to as tagging—allowed users to begin creating their own idiosyncratic metastructure for document collections. Thus, readers could view each other's annotations and even filter by reader—a major step toward what Bush called associative trails.[40]

FRESS also marked the first experiment with hypertext in the classroom. Brown offered two experimental courses—one on energy, another on poetry— in which students conducted all of their reading, writing, and postclass discussion online. Class dialogue followed the now-familiar threaded discussion format, with commentary threads attached to the related document or resource. Students wrote three times as much as in off-line classes and received better grades than their peers in a control group; class participation proved much higher than in physical classrooms, as quiet wallflowers felt comfortable piping up in the online arena. After the classes wound down, the system maintained the hypertext "corpus" of information that each class had created.[41]

The early Brown experiments proved enormously influential, allowing nonspecialist users to read and write documents online without understanding how to program a computer. Although including the term *hypertext* in the name may have been a stretch—as Nelson points out, the system had few truly hypertext capabilities beyond document linking—it marked an important milestone in the history of computing. And if nothing else, it lent an institutional imprimatur to the possibilities of hypertext in the academy.

In the years leading up to the emergence of the web, Brown's early work on HES and FRESS would provide the foundation for an even more ambitious effort, IRIS Intermedia, which was the most fully fledged hypertext environment to emerge before the web, indeed a work whose collaborative and interactive authoring capabilities have yet to be equaled by the web. The Institute for Research on Information (IRIS) was founded in 1983 with grants from the Annenberg Foundation and IBM. Its twenty-person staff began work on creating a new kind of scholarly computing environment dubbed Intermedia. Led by Norm Meyrowitz and Nicole Yankelovich, the team began work on an ambitious interactive hypertext environment.

The authoring environment provided users with a suite of interlinked tools for text editing, vector graphics, photo editing, and video and timeline editing. By allowing users to create collections of interlinked original material online, the system allowed users to create a "web" of knowledge about any topic. To keep track of the relationships between objects, Intermedia stored links in a central repository database—an approach with great advantages for ensuring the integrity of links but one that ultimately precluded the system from scaling beyond a relatively limited local deployment.

The Intermedia development team pioneered a number of innovative software features that now seem commonplace: email with hyperlinks, collaborative authoring tools, and "live" objects that could be updated dynamically across multiple applications. Most importantly, the system allowed for bidirectional links so that any document would "know" which documents pointed to it (one of the present-day web's notable shortcomings is its lack of such a backtracking feature). The system also allowed for links to point to multiple resources simultaneously using so-called link anchors. Perhaps the most important distinction between Intermedia and the modern web stemmed from Intermedia's roots as a scholarly tool: Every document in the system was editable; the system was designed as much for writing as for editing. "The tools you used to create documents were the same tools you used to link them together," recalls Yankelovich. In contrast to Intermedia's read-write tools, today's web browser works only as a reader, designed to consume rather than create.[42]

Intermedia constituted a landmark in the history of hypertext. It also proved to be ahead of its time. Despite a string of early successes, by the early 1990s the IRIS team encountered difficulties securing funding for an ambitious second-phase project to be known as the Continents of Knowledge. The National Endowment for the Humanities turned down the institute's grant request, questioning why scholars in the humanities should need computers.[43] By the time IRIS closed its doors in 1994, the institute had dwindled to a small team of three headed by its director, Paul Kahn. By this time, many key staffers had left for higher-paying jobs at private sector computer companies.

It seems a bitter irony that such a promising hypertext experiment shut its doors on the eve of the web revolution. Even though Intermedia failed to survive into the web era, the project left a trail of important scholarship. One of the earliest and most enthusiastic advocates of Intermedia was the Brown University professor George Landow, who would go on to write important theoretical texts like *Hypertext* and *Hyper/Text/Theory* that brought a critical eye to the philosophical implications of hypertext. Describing one IRIS project, a collection of materials about Alfred Tennyson's "In Memoriam," Landow described how teaching the text in a hyperlinked environment brought out its networked qualities by capturing "the multilinear organization of the poem" and describing how "using the capacities of hypertext, the web permits the reader to trace from section to section several dozen leitmotifs that thread through the poem."[44] In

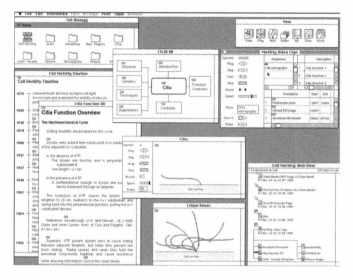

FIGURE 33. The Intermedia system, circa 1989. Courtesy of George P. Landow.

his 1992 book *Hypertext*, Landow articulated a vision that unified the literature of the nascent hypertext movement with the aims of contemporary literary scholarship. Invoking both Nelson and the theorist Jacques Derrida, Landow penned an elegant jeremiad arguing for a new model of authorship, insisting that "we must abandon conceptual systems founded upon ideas of center, margin, hierarchy, and linearity and replace them with ones of multi-linearity, nodes, links, and networks."[45] He goes on to explore the deep parallels between hypertext—with its emphasis on the decentering of the text and reduction of authority—with contemporary trends in literary theory, notably the work of Foucault and Derrida. Landow even goes so far as to argue for reconstituting our ideas of authorship: "In reducing the autonomy of the text, hypertext reduces the autonomy of the author. Lack of textual autonomy, like lack of textual centeredness, immediately reverberates through conceptions of authorship. . . . Similarly, the unboundedness of the new textuality disperses the author as well."[46]

Echoing Nelson's call to break down the rigid structures of institutional hierarchy—albeit in a more sober academic voice and from a position lodged squarely inside the academy—Landow recognizes the politics inherent even in the seemingly simple act of creating a footnote, a transaction predicated on "divisions of text that partake of fixed hierarchies of status and power." If footnotes are replaced by hyperlinks, that power relationship becomes more fluid and negotiable, and the authority of texts now derives not from the author's act of subordination but from the reader's own constantly shifting perspective. "Hypertext linking situates the present text at the center of the textual universe, thus creating a new kind of hierarchy."[47] This new democratization of texts in which the "marginal" work steps on to a level playing field with privileged texts meshed nicely with postmodern notions of "deconstruction," dissolving the illusion of the authoritative writer and creating a new teleology of literature, with readers and writers engaged on more equal footing.

Invoking Derrida's vision of a new form of hieroglyphic writing, Landow intimated that networked hypermedia authoring might provide a remedy to the linear, sequential, hierarchical model of thinking that had taken deep root in the culture in the form of the printed book. During the age of print, the relentless march toward linear sequencing of text had effectively suppressed what the hypertext scholar Gregory Ulmer calls "the pluridimensional character of symbolic thought."[48] Hypertext seemed to hold the promise of a remedy for the linear constrictions of the printed text: postmodernism made manifest.

Landow has gone on to write provocatively about the changing nature of authorship, exploring the inherent economic impetus behind authorship. As technology changes the economics of publishing, it inevitably changes the status of authors. Today, writers "exist in a cyborg relation with a complex technology that affects us more than we affect it."[49] Arguably, however, this relationship is nothing new; technologies have always shaped people's behavior at a macro level often masked by the powerful subjective sensations of individuals. Landow admits that by taking such a provocative stance, "I horrify a lot of people in literature and the humanities."[50]

Elsewhere across the academic landscape of the 1980s, other promising hypertext experiments were taking shape. In 1989, Wendy Hall and a team of researchers at the University of Southampton launched Microcosm, one of the first fully functional networked hypertext systems. A self-taught programmer with a background in mathematics, Hall had become fascinated with first-generation personal computers like the Commodore in the early 1980s, initially building instructional software for teaching colleagues about topics like electrical power generation and biological diversity. Over time, she grew interested in the question

FIGURE 34.   Wendy Hall working on Microcosm, date unknown. Courtesy of Wendy Hall.

of how to stitch together different types of electronic artifacts like photos, documents, and audio recordings. "Whatever system I was using, whether it was Word or a database or a spreadsheet or whatever, I wanted links that went across those processes, across applications. So I thought of links as being separate entities that you could apply. What if someone's written an essay or a criticism, or there's a textbook about Mountbatten? We want to link to that as well, you know. Those were the problems I was trying to solve."[51]

The first version of Microcosm allowed users to create links between documents by storing the links in a centralized database—in contrast to the World Wide Web's mechanism of embedding links in a markup language within the document.

By relying on a "linkbase" of stored metadata rather than markup language encoded in documents, Microcosm could keep track of a document's location and ensure that the links would always work. Moreover, a developer could adapt the links to be shown depending on the nature of an inbound query to improve relevance to a particular user's needs—adapting to the user's level of knowledge and incorporating new links as additional information became available. The system also continually updated itself by running text searches on every document in the system and adding new links to the linkbase along the way.

Eventually, Hall created a company around the software called Multicosm (later Active Navigation). At the time, many larger, more established software companies were starting to experiment with hypertext systems. Companies like Apple, IBM, Xerox, and Sun Microsystems were all developing systems with some degree of hypertext-like functionality for organizing documents. Indeed, 1980–90 marked the most productive period of hypertext exploration in computer science laboratories across the world. But these were all bespoke efforts with little to bind them together. In the meantime, the global Internet was expanding, raising the obvious question, "How would these competing visions of a networked world come to fruition?"

The Web That Was

Throughout the 1980s and early 1990s, the hypertext movement quietly gathered steam, spawning other early experimental environments like Apple's Hypercard, Perseus, and early Internet information-sharing applications like the Wide Area Information Server. None of these systems achieved widespread use, but each contributed to a growing wave of interest in a new humanist mode of computing that took advantage of the possibilities of hypertext.

FIGURE 35. Screenshot of Sir Tim Berners-Lee's original World Wide Web environment running on the NEXT platform. Courtesy of CERN (1991).

One application that gained an enthusiastic following in the years leading up to the web was Gopher, originally developed as a campus information system for the University of Minnesota by Mark McCahill and Farhad Anklesaria. Named for the university's mascot (and for its ability to "fetch" information), Gopher functioned as a large distributed menu system, enabling content owners to publish content in a unified browsing environment. Gopher presented users with a series of nested menus in simple lists of linear text. Each text link pointed to an object within the directory tree, which might be stored locally on the host machine or—a glimmer of the web—on a remote computer accessible over the Internet. Although the system followed a strictly hierarchical model, allowing navigation between resources only through nested menus, it nonetheless may have paved the way for the evolution of the web by setting in motion a process of linking documents across disparate geographical locations. Gopher proved enormously popular in the academic computing labs of the early 1990s and, although largely forgotten today, almost certainly played a role in seeding popular interest in the notion of remote file sharing. By 1993, a thriving Internet subculture had emerged in campus computing environments.

Paradoxically, this computing subculture would never have emerged had it not been for the US Department of Defense, which had funded development of the Internet since the late 1960s. Working under the auspices of the ARPANET project, pioneering engineers such as Vinton Cerf and Robert Kahn created distributed networking protocols like TCP/IP (Transmission Control Protocol/Internet Protocol) that would ultimately form the backbone for the modern Internet. In 1985, the National Science Foundation created NSFNet, the first national network specifically designed to support institutions of higher education. Other countries opened up similar academic networks, taking advantage of the open Internet technical standards to create a worldwide network capable of handling email and exchanging files between servers distributed around the globe.

In the years before the National Science Foundation opened the Internet up to commercial use in 1991, an enthusiastic group of dedicated hackers, university staff, and scientific researchers were busily experimenting with new tools for sharing, retrieving, and organizing information online. At the CERN (European Organization for Nuclear Research) laboratory in Switzerland, a young researcher named Tim Berners-Lee was working on a new software tool that would soon reverberate around the world.

Nelson had predicted that the future of hypertext would spring not from large institutions or top-down computing initiatives but from innovation by free-thinking individuals. Berners-Lee was about to prove him right.

As a child, Berners-Lee's family had owned a copy of a nineteenth-century English almanac titled *Enquire within upon Everything*. In sensible Victorian prose, the book contained a compendium of useful tidbits on a host of seemingly unrelated topics. As the editor of the 1875 edition wrote, "Whether you wish to model a flower in wax; to study the rules of etiquette; to serve a relish for breakfast or supper; to plan a dinner for a large party or a small one; to cure a headache; to make a will; to get married; to bury a relative; whatever you may wish to do, make, or to enjoy, provided your desire has relation to the necessities of life, I hope you will not fail to *Enquire Within*."[52] When Berners-Lee released his first version of his prototypical information-sharing tool in 1980, he decided to call it Enquire. Unfortunately for historians of the web, all traces of this early proto-web application disappeared in a calamitous hard drive crash. Undeterred, Berners-Lee started from scratch on developing a new incarnation of his hypertext information retrieval tool. In 1989, he released his program under a new name: WorldWideWeb.

The rest is not quite history.

As of January 2022, there were more than 56.5 billion web pages residing on 1.92 billion websites around the world.[53] In less than two decades, the program that took shape on Berners-Lee's desktop has circled the globe and changed the lives of a huge swath of the planet's population. As the web has worked its way into the fabric of humanity's cultural, economic, and political lives, networked information retrieval has radically reshaped the contours of our old institutional hierarchies. The web has forced dramatic transformations inside many corporations, posed challenges to the authority of the press, and completely transformed the way many people interact. Old fixed systems like library catalogs, controlled vocabularies, and manual indexes seem increasingly problematic in the digital age; new fluid systems like web search engines and collaborative filtering tools seem to be displacing them. Once again, old hierarchies are giving way to emergent networks, with long-term effects that we cannot entirely foresee.

Nelson, who arguably started it all, has disavowed the current web as an ill-conceived perversion of his original idea, a simplistic implementation of "chunking hypertext," the most rudimentary form of hypertext. Staying true to his iconoclastic form, Nelson issued a defiant polemic on his personal website, presented in unadorned ASCII text:

> I Don't Buy In
>
> The Web isn't hypertext, it's DECORATED DIRECTORIES!
>
> What we have instead is the vacuous victory of typesetters over authors, and the most trivial form of hypertext that could have been imagined.
>
> The original hypertext project, Xanadu®, has always been about pure document structures where authors and readers don't have to think about computerish structures of files and hierarchical directories. The Xanadu project has endeavored to implement a pure structure of links and facilitated re-use of content in any amounts and ways, allowing authors to concentrate on what mattered.
>
> Instead, today's nightmarish new world is controlled by "webmasters", tekkies unlikely to understand the niceties of text issues and preoccupied with the Web's exploding alphabet soup of embedded formats. XML is not an improvement but a hierarchy hamburger. Everything, everything must be forced into hierarchical templates! And the "semantic web" means that tekkie committees will decide the world's true concepts for once and for all. Enforcement is going to be another problem:) It is a very strange way of thinking, but all too many people are buying in because they think that's how it must be.
>
> There is an alternative.
>
> Markup must not be embedded. Hierarchies and files must not be part of the mental structure of documents. Links must go both ways. All these fundamental errors of the Web must be repaired. But the geeks have tried to lock the door behind them to make nothing else possible.
>
> We fight on.[54]

Nelson's angry dismissal undoubtedly springs in part from personal disappointment at the web's ascendance over his own vision for Xanadu, which now seems relegated to the status of a permanent historical footnote. Nonetheless, he correctly identifies many critical limitations of the web: the evanescence of web links, the co-opting of hypertext by corporate interests, and the emergence of a new "priesthood" of programmers and gatekeepers behind the scenes who still exert control over the technological levers powering the commercial web.

Even the web's own inventor, Berners-Lee, has written despondently over its current state. Like Nelson, Berners-Lee originally envisioned a two-way authoring environment in which browsers would function as both clients for reading documents and tools for writing new ones. Such a democratic ethos ultimately runs counter to the profit motives of commercial software companies, however, which have a strong vested interest in treating users as consumers rather than creators.

For all its limitations, however, the web's colossal success serves as a ringing vindication of its forefathers' vision. Even though it may be far from perfect, the web has proved to be an enormously important democratizing technology. It is a profoundly humanist medium, open, accessible, and largely populated by non-programmers who have found ways to express themselves using a panoply of available technologies. The explosion of blogging, photo sharing, and other forms of personal expression testify to the viability of a computing environment populated not by "priests" of the computer but by ordinary citizens.

The web's popularity has proved a mixed blessing for the original hypertext visionaries. Although its success has vindicated their vision, it has also effectively limited the research horizons for many otherwise promising paths of inquiry. Whereas the two decades before the web saw a flurry of innovation and hopeful experimentation in networked information systems, the current dominance of web standards has created a technological monolith that seems to preclude further experimentation.

As the web continues to grow, however, its structural weaknesses are coming into stark relief. The sheer fluidity of web pages makes it all but impossible to create fixed reference points; the lack of archiving functions leaves much of the web in a state of perpetual amnesia; hyperlinks work in only one direction, making it difficult for the owner of a document to identify incoming links; and browsers are still designed primarily as tools of consumption, not creativity. At a more visceral level, the web still carries the legacy of print: rectangular "pages" remain the dominant two-dimensional construct that seems woefully inadequate to support the range of possible interactions that networked computers could support. The Internet theorist McKenzie Wark has suggested that we need "a new spatial architecture for dealing with text," echoing the sentiments of many earlier hypertext theorists.[55] Perhaps our centuries of reliance on the familiar form of the printed book will make the page metaphor difficult to dislodge, or perhaps something new will yet emerge to break the conceptual stranglehold of a basic form that has remained essentially unchanged since the age of papyrus.

Twenty years after Gutenberg invented his printing press, a bare handful of people in Germany and France had ever seen a printed book. Less than twenty years after its invention, the World Wide Web has touched billions.

MEMORIES OF THE FUTURE

> And new philosophy calls all in doubt,
> The element of fire is quite put out;
> The sun is lost, and the earth, and no man's wit
> Can well direct him where to look for it.
> And freely men confess that this world's spent,
> When in the planets, and the firmament
> They seek so many new; they see that this
> Is crumbled out again to his atomies
> Tis all in pieces, all coherence gone;
> All just supply, and all relation.
>
> —John Donne, "An Anatomy of the World"

Today, we stand at a precipice: between the near-limitless capacity of computer networks and the real physical limits of human comprehension. More than four billion people have now used the Internet to access information. Millions of children in developing countries use laptops costing less than $100 to connect online.[1] In an era when almost anyone can publish, digital technology has fueled an unprecedented surge of individual expression. In the legions of social media posters, bloggers, artists, politicians, journalists, and everyday citizens taking advantage of the Internet's ease of communication, we are witnessing the rise of vast populist networks threatening the power of old institutional hierarchies. The success of pop sociology books like Malcolm Gladwell's *The Tipping Point* and James Surowiecki's *The Wisdom of Crowds* seems to attest further to the appeal of the new populism, suggesting a growing faith in an emergent order guided not by an elite class of philosophers but by the will of the people.

In the sixteen years since *Glut* first appeared in 2007, the explosive growth of social media platforms has ushered in an unprecedented era of human creativity and knowledge production—as well as a host of unintended consequences, changing social norms, and wide-scale economic and political disruptions—misinformation, surveillance capitalism, and the rise of a new political tribalism, to name a few. Trust in institutions of all stripes seems to be withering, as people increasingly place their trust in self-organizing networks over top-down hierarchical systems that emanate from institutions. This is an oversimplification, of course, as most people continue to engage with both kinds of systems. But at the moment, it certainly seems that in the age-old dance between networks and hierarchies, networks currently enjoy the upper hand.

If the history of information systems teaches us anything, it is that hierarchies continually emerge out of networks and vice versa. Across the Internet, new communities are constantly emerging, dissolving, and reforming—a phenomenon now massively accelerated by the proliferation of social media. This burgeoning network activity does not spell the end of institutional hierarchies, however. Take, for example, Wikipedia, the ambitious public Internet encyclopedia that has grown to encompass millions of entries written by thousands of individual users. For all its populist appeal, Wikipedia's sheer growth forced it to develop a set of hierarchical control systems, instituting a governing organization and approval processes to exert a measure of top-down regulation over the sprawling bottom-up activity of its contributors. As the network has grown, a new governing hierarchy has emerged.

In a similar vein, the surfeit of networked data across the open Internet has given rise to the emergence of a handful of centralized hierarchical organizations— so-called Big Tech companies—that channel the flow of that information. But this state of affairs seems unlikely to persist forever. Over the long term, further creative disruption seems all but inevitable. For now, though, we find ourselves navigating a new information ecosystem that in many ways seems to fulfill Walter Ong's prediction of a coming era of secondary orality: a postliterate mode of being in which people move beyond the "individualized introversion" of reading and writing toward more collaborative, socially constructed ways of knowing.[2]

This transition is so widespread and so interconnected with our underlying social, cultural, and political systems that it will likely take decades (at least) before anyone can fully evaluate the long-term impact of these shifts. The recent spate of hyperbolic negative press coverage about the dangers of social media notwithstanding, our networked age has also delivered considerable benefits to humanity at large: unprecedented ease of access to information, far exceeding even the most utopian visions of H. G. Wells, Vannevar Bush, and Paul Otlet; the ability to commune and collaborate across great distances, as evidenced by the transition so many office workers made to virtual work during the COVID-19 pandemic; millions of people lifted out of poverty in the developing world via social commerce; and the exuberant creativity that is so evident in short-form videos created and posted by everyday people on TikTok and other social platforms. As the technology historian Melvin Kranzberg put it, "Technology is neither good nor bad; nor is it neutral."[3]

Whether the most idealistic visions of a digital utopia ever come to fruition, we are already witnessing social, political, and cultural change on a scale comparable only to the greatest previous axial periods in the history of technology. During the Ice Age information explosion, the new technologies of symbolism allowed previously loose-knit social groups to form larger tribal organizations. In ancient Sumer, the advent of writing allowed traders to broaden their circles of trust and pave the way for the first bureaucracies. In medieval Europe, the continent's great descent into illiteracy and the subsequent renewal of oral culture provided the catalyst for the literary innovations of the medieval scriptoria. Centuries later, the violent disruptions of the Gutenberg era brought mass literacy into conflict with the older folk traditions that had predominated outside the cloisters, resulting in a series of transformations that still reverberate today.

With the rise of the Internet, we may be witnessing a similar phenomenon taking shape today. Many of our familiar digital communication tools like email, instant messaging, and social media may seem, on the surface, like just another form of writing. But if we look closely at how people use these tools, we can see that many of these new modes of communication bear at least as much resemblance to the spoken word as to traditional forms of writing. The conversational tone of email, the pidgin-like dialects of chat and asynchronous communication channels, or the evolving language of gifs and emojis all seem to echo the casual, immediate rhythms of speech. Of course, large parts of the Internet still consist of traditional literary artifacts like books, newspapers, essays, dissertations, and so forth. But as more and more people get online, their interactions feel increasingly like a vast, far-flung conversation. Ultimately, the Internet's popularity may have less to do with a renewed public love of reading and writing than with our deep-seated need to talk to each other.

As Steven Pinker points out, most children learn to write only with great difficulty; but we are born babblers, the "talking ape."[4] Until recently, most people on earth remained illiterate (and today, more than one in five people still cannot read).[5] But almost every one of us is a natural talker. So perhaps it should come as no surprise that our most durable information systems—mythologies, religions, and various forms of practical know-how—have persisted not by dint of libraries, indexes, or taxonomic systems but largely through that timeless and durable mode of knowing: word of mouth. Today, we may be witnessing the latest manifestation of that ancient disposition toward spoken language, with the reemergence in electronic form of oral patterns that have been hiding in plain sight for generations.

Perhaps it should come as no surprise, then, that so many people have taken to the Internet, an artificial technology that lends itself to both "oral" and literate forms of expression. Often, both modes coexist with each other. To take a familiar example, many commercial websites, like Amazon.com, offer independent reviews to accompany their product listings. Typically, these come in two flavors: "expert" reviews, usually licensed from publications, and customer-written reviews contributed by users. Frequently, these reviews appear side by side, yet separately. Customer reviews seem to fit Ong's criteria of an oral medium; they acquire value additively and in aggregate—that is, we look at the cumulative customer ratings for a product more than an individual review; they are empathetic—usually written in the first person—and participatory, insofar as they invite conversation. Editorial reviews, by contrast, bear the hallmarks of literate culture: they are subordinative—that is, presuming a voice of authoritative judgment—analytic and objectively distanced, and abstract (usually written in the third person). These two forms—the oral and the written—coexist, yet they never quite meet. Most readers readily make the distinction between the two forms, and they may award more authority to the individual editorial review over the individual customer review. But the cumulative "oral tradition" of the customer reviews as a whole may often carry more weight than the individual editorial review.

The distinction between oral and literate cultures surfaces not just in the editorial content of websites but increasingly at deeper structural levels as well. Many websites now support "social tagging" or other user-driven mechanisms for allowing visitors to organize the contents of a website. Popular sites like YouTube, Flickr, and de.licio.us give users a measure of control over how they organize the contents of the sites, allowing users to "tag" entries with keywords of their own choosing. The result is an emergent classification—sometimes called a folksonomy (a neologism coined by Thomas Vanderwal)—that, although typically flawed, allows users a great deal of flexibility and control over the information at hand.[6] Over time, users typically reach a rough consensus over how a particular object should be categorized, and an emergent structure begins to take shape. A few sites, like Amazon, have tried to reconcile these bottom-up systems with more traditional top-down organizational models with mixed results. The tension between oral and literate cultures seems to resist a smooth reconciliation.

At the level of organizational dynamics, the tension between literacy and orality manifests as a polarity between "fixity" and "fluidity" (to borrow John Seely Brown and Paul Duguid's terms). Examples of fixed systems include books, filing systems, taxonomies, and ontologies: the static instruments of literacy. Fluid systems include email, instant messaging, and chat rooms: the shifting miasma of the new orality. Fixed systems beget hierarchies, while fluid systems lend themselves to the social motility of networks. In recent years, many corporations have embraced the mantra of flattening their organization charts to become more responsive in the face of technological change. Even such hidebound bureaucracies as the US Army are working to transform themselves into "edge organizations," driving decision-making responsibility away from the old command-and-control centers and toward the outer rim of the organization in an effort to enable individuals to take better advantage of new communications technology. In the process, organizations large and small are undergoing radical and disruptive change.[7]

Fixity and fluidity can and do coexist, however. "While it's clear that self-organization is extraordinarily productive, so too is formal organization," write Brown and Duguid. "Indeed, the two perform an intricate (and dynamic) balancing act, each compensating for the other's failings. Self-organization overcomes formal organizing's rigidity. Formal organization keeps at bay self-organization's tendency to self-destruct."[8] In other words, hierarchies and networks do not necessarily have to stand in opposition; they may not only coexist but ultimately prove consilient. As Kevin Kelly puts it in his 1994 book *Out of Control*:

> In the human management of distributed control, hierarchies of a certain type will proliferate rather than diminish. That goes especially for distributed systems involving human nodes—such as huge global computer networks. Many computer activists preach a new era in the network economy, an era built around computer peer-to-peer networks, a time when rigid patriarchal networks will wither away. They are right and wrong. While authoritarian "top-down" hierarchies will retreat, no distributed system can survive long without nested hierarchies of lateral "bottom-up" control. As influence flows peer to peer, it coheres into a chunk—a whole organelle—which then becomes the bottom unit in a larger web of slower actions. Over time a multi-level organization forms around the percolating-up control: fast at the bottom, slow at the top.[9]

Although the tension between hierarchies and networks may in the long run prove complementary, in the short term the results can look a lot like chaos. Today, the old hierarchies of the literate world are reeling from a collision with the chaotic energies of oral culture. This is more than just a clash of epistemologies; it is a conflict with potentially revolutionary implications. Throughout most of human history, the top layers of the institutional hierarchy have always belonged to the literate elite. Without a lettered class there would have been no Sumerian temples, no library at Alexandria, no monastic scriptoria. Even after the Gutenberg revolution, only a tiny handful of people actually produced written documents, let alone whole books. Today, however, the planet has more readers and writers than ever before, and increasingly they are connected to each other over the Internet. The long-term results of such a dramatic technological shift may be impossible to predict, but in the near term one thing seems plain enough: as people find their way online, they seem to coalesce into small groups.

Before the age of television, many historians believed that the spread of literacy signaled the forward march of technological progress in which human civilization was moving inexorably forward toward higher degrees of social complexity. In this view, the spread of information technologies helped smaller communities coalesce into larger entities: bands into tribes, tribes into city-states, city-states into nations, nations into empires, and so on. Culture and technology propelled each other forward. Recent history, however, seems to support an alternative view: that in our modern technological era, human culture may not be moving unidirectionally at all but rather multidirectionally. The notion of inevitable progress toward hierarchical complexity began to fracture in the 1960s with the rise of the great modern liberation movements, especially in the United States: the civil rights movement in the American South, the anti-war movement, feminism, sexual liberation, gay rights. Although all of these reform movements had deeper historical roots that stretched much further back into American history, they coalesced in the 1960s in no small part due to the proliferation of electronic media. Suddenly, women, minorities, homosexuals, and other previously marginalized groups had access to technologies that provided new windows into the culture at large. Televised images provided a unifying reference point that encouraged like-minded individuals to seek each other out. Small self-organized communities emerged around common causes and shared values. In each of these cases, spontaneous networks emerged as individuals began to communicate with each other, thanks in large part to the facilitative power of electronic media. As these spontaneous networks coalesced, they began to challenge the existing institutional power structures. Black citizens demanded the vote; women demanded equal rights; homosexuals demanded legal equality. The same dynamic played out in smaller groups as well, as the cultural convulsions of the 1960s brought wave after wave of spontaneous bottom-up challenges to the traditional top-down establishment. The success of these self-organized movements seemed to call into question the unidirectional progression from smaller to large units of social organization. Instead, it seemed, the age of electronic media was leading not to the forward march of civilization but rather to social and cultural fragmentation. That trajectory toward social atomization would accelerate further with the advent of the personal computer and, later, the Internet.[10]

The 1960s liberation movements provided the social backdrop for the emergence of the personal computer industry. The first-generation PC hackers, many of whom traced their lineage to Doug Engelbart's 1968 San Francisco demonstration (see chapter 11), aligned themselves with the countercultural values of the hippie movement, embracing an ethos of personal empowerment in contrast to the old top-down, command-and-control systems of the mainstream military-industrial computing industry.[11] Ted Nelson, the philosophical godfather of the web, always portrayed himself as an ardent antiestablishmentarian. For the early programmers, the PC represented a vehicle for personal liberation from the oppressive control of hierarchical institutions.

As PCs moved into the workplace in the 1980s, they triggered a struggle between self-directed "power users," who wanted to take advantage of their increasingly powerful desktop computers, and old-guard information technology departments that continued to insist on centralized control (a tension that still carries over to this day in many large organizations). That simmering tension would finally boil over when the Internet started allowing those users to network with each other directly, bypassing the old command-and-control systems. For many old-guard organizations, the results proved nothing short of paroxysmic. As the Internet gave employees more avenues for reaching out to each other—via the web, email, or instant messaging—they started to coalesce into spontaneous self-organizing groups that often bore little resemblance to the formal structure of the organization chart. Decades after the liberation movements of the 1960s had peaked, the same revolutionary ethic was echoing in the corridors of corporate power. Organizations found themselves forced to change in response to the emerging technology. Fueled by the growth of personal computing and network technology, many organizations have since had to come to terms with an ongoing transfer of power, away from the old central planning hierarchies and toward networks of people choosing to affiliate with each other online.

Although this tendency toward self-determination might seem like an effect of the Internet's democratic architecture, such behavior also harkens back to our deep-rooted social instincts. For most of our species' history, human beings have interacted in small tightly woven communities: families, villages, guilds, and other social groups whose members were bound by ties of direct kinship or close personal affiliation. Only in the past few thousand years have people allowed themselves to be governed by institutional bodies. On the scale of evolutionary history, institutions remain a short-lived hypothesis. Yet for tens of thousands of years, human beings have interacted as social animals, following unwritten norms strengthened by kinship, reinforced by the limbic responses that strengthen our personal relationships, and transmitted through the spoken word. Today, we are seeing those instincts returning to the fore as people adapt new technologies to invoke the ancient emotional circuitry that carried us through the age before symbols. The explosive growth of networked consciousness, powered by embedded technologies that mesh to the point of vanishing into our daily lives, may augur the return of these old preliterate ways of knowing. The future of memory, then, may lie not in our heads but in our hearts.

JOHN WILKINS'S UNIVERSAL CATEGORIES

General; namely those Universal notions, whether belonging more properly to Things; called TRANSCENDENTAL

> GENERAL. I
> RELATION MIXED. II
> RELATION OF ACTION. III
> Words; DISCOURSE. IV

Special; denoting either
CREATOR. V
Creatures; namely such things as were either created or concreated by God, not excluding several of those notions, which are framed by the minds of men, considered either

> Collectively; WORLD. VI
> Distributively; according to several kinds of Beings. whether such as do belong to
> Substance;

> Inanimate; ELEMENT. VII
> Animate; considered according to their several
> Species; whether
> Vegetative

Imperfect; as Minerals,
    STONE. VIII
    METAL. IX

Perfect; as Plant,
    HERB consid. accord. to the
    LEAF. X
    FLOWER. XI.
    SEED-VESSEL. XII
    SHRUB. XIII
    TREE. XIV

Sensitive;
    EXANGUIOUS. XV
    Sanguineous;
    FISH. XVI
    BIRD. XVII
    BEAST. XVIII

Parts;
    PECULIAR. XIX
    GENERAL. XX

Accident;
Quantity;
    MAGNITUDE. XXI
    SPACE. XXII
    MEASURE. XXIII

Quality; whether
    NATURAL POWER. XXIV
    HABIT. XXV
    MANNERS. XXVI
    SENSIBLE QUALITY. XXVII
    SICKNESS. XXVIII

Action
    SPIRITUAL. XXIX
    CORPOREAL. XXX
    MOTION. XXXI
    OPERATION. XXXII

Relation; whether more Private
    OECONOMICAL. XXXIII
    POSSESSIONS. XXXIV
    PROVISIONS. XXXV

Publick
    CIVIL. XXXVI
    JUDICIAL. XXXVII
    MILITARY. XXXVIII
    NAVAL. XXXIX
    ECCLESIASTICAL. XL.

THOMAS JEFFERSON'S 1783 CATALOG OF BOOKS

Books may be classed from the Faculties of the mind, which being
I. Memory. II. Reason. III. Imagination
are applied respectively to
I. History. II. Philosophy. III. Fine Arts.

HISTORY.	CIVIL.	CIVIL PROPER.	ANTIENT [ANCIENT].	ANTIENT HIST.
			Modern.	Foreign.
				British.
				American.
		Ecclesiastical.		Ecclesiastical.
	Natu-	Physics.		Natl. Philosy.
	ral.			Agriculture.
				Chemistry.
				Surgery.
				Medecine
		Natl. histy.	Animals.	Anatomy.
		propr.		Zoology.
			Vegetables.	Botany.
			Minerals.	Mineralogy
		Occupations of Man.		Technical arts.

PHILOSOPHY.	MORAL.	ETHICS.		*MORAL PHILOSY.
				Law Nature & nations
		Jurisprudence.	Religious.	Religion.
			Municipal.	Domestic.	Equity.
				Common Law.
				L. Merchant.
				L. Maritime.
				L. Ecclesiastl.
			Foreign.	Foreign Law.
			Oeconomical.	Politics.
				Commerce.
	Mathematical.	Pure.		Arithmetic.
				Geometry
		Physico-Mathematical.		Mechanics.
				St. Statics
				[*illegible*] Dynamics
				Pneumatics.
				Phonics.
				Optics.
				Astronomy.
				Geography.

FINE ARTS.		GARDENING.		GARDENING.
		Architecture.		Architecture.
		Sculpture.		Sculpture.
		Painting.		Painting.
		Music.	Theoretical.	Music Theory.
			Practical.	Music. Vocal.
				Music. Instrumentl.
		Poetry.	Narrative.	Epic.
				Romance.
			Dramatic.	Tragedy.
				Comedy.
				Pastorals.
				Odes.
				Elegies.
				Dialogue.
			Didactic.	Satire.
				Epigram.
				Epistles.
		Oratory.		Logic.
				Rhetoric.
				Oration.
		Criticism.		Criticism.
Authors who have written in various branches.				Polygraphical.

THE DEWEY DECIMAL SYSTEM

0 GENERAL

10 Bibliography

20 Book Rarities

30 General Cyclopedias

40 Polygraphy

50 General Periodicals

60 General Societies

70

80

90

100 PHILOSOPHY

110 Metaphysics

120

130 Anthropology

140 Schools of Psychology

150 Mental Faculties

160 Logic

170 Ethics

180 Ancient Philosophies

190 Modern Philosophies

200 THEOLOGY

210 Natural Theology

220 Bible

230 Doctrinal Theology

240 Practical and Devotional

0 GENERAL

250 Homiletical and Pastoral

260 Institutions and Missions

270 Ecclesiastical History

280 Christian Sects

290 Non-Christian Religions

300 SOCIOLOGY

310 Statistics

320 Political Science

330 Political Economy

340 Law

350 Administration

360 Associations and Institutions

370 Education

380 Commerce and Communication

390 Customs and Costumes

400 PHILOLOGY

410 Comparative

420 English

430 German

440 French

450 Italian

460 Spanish

470 Latin

480 Greek

490 Other Languages
500 NATURAL SCIENCE
510 Mathematics
520 Astronomy
530 Physics
540 Chemistry
550 Geology
560 Paleontology
570 Biology
580 Botany
590 Zoology
600 USEFUL ARTS
610 Medicine
620 Engineering
630 Agriculture
640 Domestic Economy
650 Communication and Commerce
660 Chemical Technology
670 Manufactures
680 Mechanic Trades
690 Building
700 FINE ARTS
710 Landscape Gardening
720 Architecture
730 Sculpture
740 Drawing and Design

500 NATURAL SCIENCE
750 Painting
760 Engraving
770 Photography
780 Music
790 Amusements
800 LITERATURE
810 Treatises and Collections
820 English
830 German
840 French
850 Italian
860 Spanish
870 Latin
880 Greek
890 Other Languages
900 HISTORY
910 Geography and Description
920 Biography
930 Ancient History
940 Modern Europe
950 Modern Asia
960 Modern Africa
970 Modern North America
980 Modern South America
990 Modern Oceanica and Polar

THE UNIVERSAL DECIMAL CLASSIFICATION

0 SCIENCE AND KNOWLEDGE

00	Prolegomena. Fundamentals of knowledge and culture. Propaedeutics
01	Bibliography and bibliographies. Catalogues
02	Librarianship
030	General reference works (as subject)
050	Serial publications, periodicals (as subject)
06	Organizations of a general nature
070	Newspapers. The Press. Journalism
08	Polygraphies. Collective works
09	Manuscripts. Rare and remarkable works

1 PHILOSOPHY. PSYCHOLOGY

101	Nature and role of philosophy
11	Metaphysics
122/129	Special Metaphysics
13	Philosophy of mind and spirit. Metaphysics of spiritual life
14	Philosophical systems and points of view
159.9	Psychology

| 16 | Logic. Epistemology. Theory of knowledge. Methodology of logic |
| 17 | Moral philosophy. Ethics. Practical philosophy |

2 RELIGION

| 2-1/-9 | Special auxiliary subdivision for religion |
| 21/29 | Religious systems. Religions and faiths |

3 SOCIAL SCIENCES

303	Methods of the social sciences
304	Social questions. Social practice. Cultural practice. Way of life.
305	Gender studies
306	Sociography. Descriptive studies of society
311	Statistics as a science. Statistical theory
314/316	Society
32	Politics
33	Economics. Economic science
34	Law. Jurisprudence
35	Public administration. Government. Military affairs
36	Safeguarding the mental and material necessities of life
37	Education
39	Cultural anthropology. Ethnography. Customs. Manners. Traditions.

4 (UNUSED)

5 MATHEMATICS. NATURAL SCIENCES

502/504	Environmental science. Conservation of natural resources
51	Mathematics
52	Astronomy. Astrophysics. Space research. Geodesy
53	Physics
54	Chemistry. Crystallography. Mineralogy
55	Earth sciences. Geological sciences
56	Palaeontology

57	Biological sciences in general
58	Botany
59	Zoology

6 APPLIED SCIENCES. MEDICINE. TECHNOLOGY

60	Biotechnology
61	Medical sciences
62	Engineering. Technology in general
63	Agriculture and related sciences and techniques
64	Home economics. Domestic science. Housekeeping
65	Communication and transport industries. Accountancy. Business management
66	Chemical technology. Chemical and related industries
67	Various industries, trades and crafts
68	Industries, crafts and trades for finished or assembled articles
69	Building (construction) trade. Building materials. Building practice and procedure

7 THE ARTS. RECREATION. ENTERTAINMENT. SPORT

7.01/.09	Special auxiliary subdivision for the arts
71	Physical planning. Regional, town and country planning. Landscapes, parks, gardens
72	Architecture
73	Plastic arts
74	Drawing. Design. Applied arts and crafts
75	Painting
76	Graphic art, printmaking. Graphics
77	Photography and similar processes
78	Music
79	Recreation. Entertainment. Games. Sport

8 LANGUAGE. LINGUISTICS. LITERATURE

80	General questions relating to both linguistics and literature. Philology
81	Linguistics and languages
82	Literature

9 GEOGRAPHY. BIOGRAPHY. HISTORY

902/908	Archaeology. Prehistory. Cultural remains. Area studies
91	Geography. Exploration of the Earth and of individual countries. Travel. Regional geography
92	Biographical studies. Genealogy. Heraldry. Flags
93/94	History

COMMON AUXILIARY SIGNS

+ Coordination. Addition (plus sign)
/ Consecutive extension (oblique stroke sign)
: Simple relation (colon sign)
:: Order-fixing (double colon sign)
[] Subgrouping (square brackets)
* Introduces non-UDC notation (asterisk)
   A/Z Direct alphabetical specification

COMMON AUXILIARY NUMBERS

= . . . Common auxiliaries of language
01/08 Special auxiliary subdivision for origins, periods of language and phases of development
276/282 Special auxiliary subdivision for language usage, dialects and variants
=00 Multilingual. Polyglot
=030 Translated documents. Translations
=1/=9 Languages (natural and artificial)
(0 . . .) Common auxiliaries of form
(0.02/.08) Special auxiliary subdivisions for document form
(01) Bibliographies

(02) Books in general

(03) Reference works

(04) Non-serial separates. Separata

(05) Serial publications. Periodicals

(06) Documents relating to societies, associations, organizations

(07) Documents for instruction, teaching, study, training

(08) Collected and polygraphic works. Forms. Lists. Illustrations. Business publications

(09) Presentation in historical form. Legal and historical sources

(1/9) Common auxiliaries of place

(1) Place and space in general. Localization. Orientation

(1-0/-9) Special auxiliary subdivision for boundaries and spatial forms of various kinds

(2) Physiographic designation

(3/9) Individual places of the ancient and modern world

(= ...) Common auxiliaries of human ancestry, ethnic grouping, and nationality

(=01) Human ancestry groups

(=1/=8) Linguistic-cultural groups, ethnic groups, peoples

(=1:1/9) Peoples associated with particular places

"..." Common auxiliaries of time

"0/2" Dates and ranges of time (CE or AD) in conventional Christian (Gregorian) reckoning

"3/7" Time divisions other than dates in Christian (Gregorian) reckoning

-0 ... Common auxiliaries of general characteristics: Properties, Materials, Relationships

-02 Common auxiliaries of properties

-03 Common auxiliaries of materials

-04 Common auxiliaries of relations, processes and operations

-05 Common auxiliaries of persons and personal characteristics

RANGANATHAN'S COLON CLASSIFICATION

FACETS

Colon Classification uses five primary categories, or facets, called *PMEST*:

, Personality, the most specific or focal subject.

; Matter or property, the substance, properties or materials of the subject.

: Energy, including the processes, operations and activities.

. Space, which relates to the geographic location of the subject.

' Time, which refers to the dates or seasons of the subject.

CLASSES

z Generalia
[material], [kind], . . .
1 Universe of Knowledge
2 Library Science
[library]; [material]: [problem]
3 Book science
4 Journalism
A Natural science
B Mathematics
    B1 Arithmetic
    B2 Algebra

B3 Analysis
B4 Other methods
B6 Geometry [space]: [method]
B7 Mechanics [matter]: [problem]
B8 Physico-mathematics
B9 Astronomy [body]: [problem]
C Physics
   C1 Fundamentals
   C2 Properties of matter [state]: [problem]
   C3 Sound [wave length]: [problem]
   C4 Heat [state]: [problem]
   C5 Light, radiation [wave length]: [problem]
   C6 Electricity [electricity]: [problem]
   C7 Magnetism [magnetism]: [problem]
   C8 Cosmic hypotheses
D Engineering
   [work], [part]; [material]: [problem]
E Chemistry
   [substance], [combination]: [problem]
F Technology
   [substance]: [problem and process]
G Biology
   [organ and special grouping]: [problem]
H Geology
   [substance]: [problem]
I Botany
   [natural group], [organ]: [problem]
J Agriculture
   [plant], [utility/part], [organ]: [problem], [nature of soil/substance/cause/
   material]: [operation]
K Zoology
   [same formula as I Botany]
L Medicine
   [organ]: [problem], [cause]: [handling]
M Useful arts
   [material]: [work]
Δ Spiritual experience and mysticism
   [religion], [entity]: [problem]
N Fine arts
NA Architecture [style], [utility], [part]: [technique]

ND Sculpture [style], [figure]; [material]: [technique]
NN Engraving
NQ Painting [style], [figure]; [material]: [technique]
NR Music [style], [music]; [instrument]: [technique]
O Literature
    [language], [form], [author], [work]
P Linguistics
    [language], [variant-stage], [element]: [problem]
Q Religion
    [religion]: [problem]
R Philosophy
    R1 Logic
    R2 Epistemology
    R3 Metaphysics [view], [subject]
    R4 Ethics [topic], [controlling principle]
    R5 Aesthetics
    R6 Favoured system 1: e.g. Indian philosophy [system], [canonical/basic text]
S Psychology
    [entity]: [problem]
T Education
    [educand]: [problem], [subject], [method]
U Geography
    [geography]
V History
    [community], [part]: [problem]
W Political science
    [type of state], [part]: [problem]
X Economics
    [business]: [problem]
Y Sociology
    [group]: [problem]: [secondary problem]
Z Law
    [community], [law 1], [law 2]

Acknowledgments

In a medieval scriptorium, it would have been considered the height of arrogance for a scribe to put his name on the cover of a book. Making books was properly understood to be a collaborative effort, involving the combined talents of scholars, scribes, binders, and artists to create a final manuscript. Although the making of books has long since evolved from a monastic craft to an industrial enterprise—and although we may now agree on the convenient fiction of an "author"—a book is still the product of many minds.

This book would not exist in its present form without the efforts of my original editor Jeff Robbins, my agent Jeanne Fredericks, and the able team at the National Academies Press. This book is also immeasurably better for the thoughtful criticism of several people who commented on various draft versions of the manuscript. For their intellectual generosity, I am especially indebted to Whit Andrews, Brent Berlin, Ed Costello, Eugene Garfield, Paul Kahn, George P. Landow, Rhodri Lewis, Peter Merholz, Thomas J. Misa, Mike Myers, Dave Pell, Nadav Savio, Rosemary Michelle Simpson, Andries van Dam, E. O. Wilson, and Nicole Yankelovich. Portions of chapter 11 first appeared as "Forgotten Forefather: Paul Otlet," *Boxes and Arrows*, November 10, 2003, https://boxesandarrows.com/forgotten-forefather -paul-otlet/.

For *Informatica*, I am indebted to Jim Lance, Susan Specter, Anne Jones, Clare Jones, Mia Renaud, and the entire team at Cornell University Press for their guidance in shepherding the manuscript to its latest incarnation. Thanks also to Shannon Mattern for her insightful suggestions on how to update and modernize the book, and to Paula Clarke Bain for conducting a master class in book indexing. I am also grateful to a number of readers who have reached out over the years to share their feedback, criticisms, and suggestions on new topics to explore. For their good company and hospitality, thanks also to Ann and Preston Browning at Wellspring House, Judy DeMocker, Jennifer Kilian, Bill and Jocelyn Powning, Chris and Kassia Scott, and John Woodward.

For their support and forbearance in granting me the time to complete these revisions, thanks to my wife, Maaike, and my sons, Colin and Elliot. Finally, my greatest thanks go to my late parents, Sarah Bird Wright and Lewis Wright, for their love and encouragement.

Notes

PREFACE

1. Tom Standage, *Writing on the Wall: Social Media, the First 2,000 Years* (New York: Bloomsbury, 2013); Markus Krajewski, *Paper Machines: About Cards & Catalogs, 1548–1929* (Cambridge, MA: MIT Press, 2011); Colin B. Burke, *Information and Intrigue: From Index Cards to Dewey Decimals to Alger Hiss* (Cambridge, MA: MIT Press, 2014).

2. Alex Wright, *Cataloging the World: Paul Otlet and the Birth of the Information Age* (New York: Oxford University Press, 2014).

INTRODUCTION

1. Tom Wolfe, *The Electric Kool-Aid Acid Test* (New York: Picador, 2008).

2. Tom Wolfe, "Digibabble, Fairy Dust and the Human Anthill," *Forbes ASAP* 164, no. 8 (1999): 212.

3. Steven Johnson, *Interface Culture: How New Technology Transforms the Way We Create and Communicate* (San Francisco: HarperEdge, 1997), 240.

4. Danny Hillis, "The Big Picture," *Wired* 6, no. 1 (January 1998), http://www.wired.com/wired/archive/6.01/hillis.html.

5. Ray Kurzweil, "The Law of Accelerating Returns," https://www.kurzweilai.net/the-law-of-accelerating-returns

6. H. G. Wells, *World Brain* (Cambridge, MA: MIT Press, 2021).

7. Pierre Teilhard de Chardin, *The Future of Man*, trans. Norman Denny (New York: Image Books/Doubleday, 2004), 162.

8. Sven Birkerts, *The Gutenberg Elegies: The Fate of Reading in an Electronic Age* (New York: Faber & Faber, 2006), 130–31.

9. Werner Künzel, quoted in Geert Lovink, "The Archaeology of Computer Assemblage," *Mediamatic* 7, no. 1 (January 1, 1992), http://www.mediamatic.net/article-8664-en.html.

10. Birkerts, *The Gutenberg Elegies*, 123.

1. NETWORKS AND HIERARCHIES

1. Émile Durkheim, Marcel Mauss, and Rodney Needham, *Primitive Classification* (Chicago: University of Chicago Press, 1963), 43–44.

2. Richard Saul Wurman, *Information Architects* (Graphis, 1997), 15.

3. 5,000,000,000,000,000,000 bytes.

4. Peter Lyman and Hal R. Varian, "How Much Information," 2003, http://www.sims.berkeley.edu/how-much-info-2003.

5. 175,000,000,000,000,000,000,000 bytes.

6. David Reinsel, John Gantz, and John Rydning, *The Digitization of the World: From Edge to Core* (Framingham, MA: IDC, November 2018), https://www.seagate.com/files/www-content/our-story/trends/files/idc-seagate-dataage-whitepaper.pdf.

7. Christopher Locke et al., *The Cluetrain Manifesto*, http://www.cluetrain.com/.

8. Manuel Castells, *The Rise of Network Society* (Chichester, UK: Wiley, 2009), 508.

9. Jeff Hawkins with Sandra Blakeslee, *On Intelligence: How a New Understanding of the Brain Will Lead to the Creation of Truly Intelligent Machines* (New York: Owl Books, 2004).

10. Nicholas Christakis and James Fowler, *Connected* (New York: Little, Brown Spark, 2009), 239.

11. Niall Ferguson, *The Square and the Tower: Networks and Power, from the Freemasons to Facebook* (New York: Penguin, 2017), 39.

12. Howard K. Bloom, *Global Brain: The Evolution of Mass Mind from the Big Bang to the 21st Century* (New York: Wiley, 2000).

13. Bloom, *Global Brain*, 31.

14. E. O. Wilson initially proposed an alternate neologism, "culturgen," but has since embraced the now overwhelmingly popular "meme." Wilson argues in favor of a narrow definition of meme, however, as a unit of "semantic memory" available exclusively to humans. Other biologists have embraced a wider definition that includes behavioral "implicit memes"—roughly, social customs—stretching as far back as 720-million-year-old prehistoric clams. Other scientists have sought a middle ground, suggesting the presence of memes in smaller numbers of species, including hominids, dolphins, and a few other mammals.

15. Bloom, *Global Brain*, 31.

16. Kevin Kelly, *Out of Control: The New Biology of Machines, Social Systems and the Economic World* (Reading, MA: Perseus Books, 1995), 12.

17. Wilson prefers the term *sematectonic* over *stigmergy*, arguing that the latter is too clumsy a term.

18. McClamrock, quoted in Andy Clark, *Being There: Putting Brain, Body, and World Together Again* (Cambridge, MA: MIT Press, 1997), 201.

19. Robert Wright, *Nonzero: The Logic of Human Destiny* (New York: Vintage, 2001), 292.

20. Carel van Schaik, "Why Are Some Animals So Smart?" *Scientific American*, April 2006, http://www.sciam.com/article.cfm?articleID=000C1E5D-B9BA-1422-B9BA83414B7F0103&ref=sciam&chanID=sa006.

21. Van Schaik, "Why Are Some Animals So Smart?"

22. Stanley I. Greenspan and Stuart G. Shanker, *The First Idea: How Symbols, Language, and Intelligence Evolved from Our Primate Ancestors to Modern Humans* (Lebanon, IN: Da Capo, 2004), 92.

23. Greenspan and Shanker, *The First Idea*, 340–41.

2. FAMILY TREES AND THE TREE OF LIFE

1. Stephen Jay Gould, foreword to Lynn Margulis and Karlene V. Schwartz, *Five Kingdoms: An Illustrated Guide to the Phyla of Life on Earth*, 3rd ed. (New York: Freeman, 1998).

2. Brent Berlin, personal correspondence with author, March 30, 2006.

3. Brent Berlin, *Ethnobiological Classification: Principles of Categorization of Plants and Animals in Traditional Societies* (Princeton, NJ: Princeton University Press, 1992), 6.

4. Berlin, *Ethnobiological Classification*, 7.

5. Conklin, quoted in Berlin, *Ethnobiological Classification*, 13.

6. Roger Brown, "How Shall a Thing Be Called?" *Psychology Review* 65 (1958): 14–21.

7. Rosch, Eleanor, as quoted in George Lakoff, *Women, Fire, and Dangerous Things: What Categories Reveal about the Mind* (Chicago: University of Chicago Press, 1987), 44.

8. Lakoff, *Women, Fire, and Dangerous Things*, 8.

9. Marcel Mauss and Emile Durkheim, *Primitive Classification* (Chicago: University of Chicago Press, 1963), 14.

10. Michael E. Hobart and Zachary Sayre Schiffman, *Information Ages: Literacy, Numeracy, and the Computer Revolution* (Baltimore: Johns Hopkins University Press, 1998), 71.

11. Durkheim and Mauss, *Primitive Classification*, 82–83.

12. Durkheim and Mauss, *Primitive Classification*, 85–86.

13. Brent Berlin, personal correspondence with author, March 30, 2006.

14. Durkheim and Mauss, *Primitive Classification*, 69.

15. Durkheim and Mauss, *Primitive Classification*, 73.

16. Durkheim and Mauss, *Primitive Classification*, 78.

17. Durkheim and Mauss, *Primitive Classification*, 79.

3. THE ICE AGE INFORMATION EXPLOSION

1. Louis Baudin, *A Socialist Empire: The Incas of Peru* (Princeton, NJ: Van Nostrand, 1961).

2. John McCrone, *Going Inside: A Tour Round a Single Moment of Consciousness* (London: Faber and Faber, 2000).

3. Steven Kuhn et al., "Ornaments of the Earliest Upper Paleolithic," *Proceedings of the National Academy of Sciences* 98, no. 3 (2001): 7641–46.

4. Clive Gamble, "Palaeolithic Society and the Release from Proximity," *World Archaeology*, 29, 3 (1998): 426–49.

5. Randall White, as quoted in William F. Allman, *The Stone Age Present: How Evolution Has Shaped Modern Life: From Sex, Violence, and Language to Emotions, Morals, and Communities* (New York: Simon & Schuster, 1994), 209.

6. Allman, *The Stone Age Present*, 215.

7. Allman, *The Stone Age Present*, 209.

8. R. Dale Guthrie, *The Nature of Paleolithic Art* (Chicago: University of Chicago Press, 2005).

9. Edward O. Wilson, *Consilience: The Unity of Knowledge* (New York: Vintage, 1999), 142.

4. THE AGE OF ALPHABETS

1. Thomas Cahill, *The Gifts of the Jews: How a Tribe of Desert Nomads Changed the Way Everyone Thinks and Feels* (New York: Nan A. Talese, 1998), 11.

2. Although Sumerians pioneered the notion of keeping time, they did not understand time the same way we do. Sumerian "hours" varied by season, depending on the availability of sunlight (a Sumerian hour was about one-sixth of a day's sunlight). The Babylonians later introduced the notion of twenty-four uniform hours in a day.

3. Vere Gordon Childe, *Man Makes Himself* (New York: New American Library, 1951).

4. Wayne A. Wiegand and Donald G. Davis, eds., *Encyclopedia of Library History* (New York: Garland, 1994), 24.

5. Paul Rincon, "'Earliest Writing' found in China," BBC News, April 17, 2003, http://news.bbc.co.uk/2/hi/science/nature/2956925.stm.

6. Walter J. Ong, *Orality and Literacy: The Technologizing of the Word* (London: Methuen, 1982), 94.

7. Ong, *Orality and Literacy*, 9.

8. Ong, *Orality and Literacy*, 87.

9. Anonymous, quoted in Lionel Casson, *Libraries in the Ancient World* (New Haven, CT: Yale University Press, 2001), 5.

10. Dewey was a leading light in the US public library movement of the nineteenth century and inventor of the Dewey Decimal System, still used in most US public libraries (see chapter 10).

11. Michael H. Harris, *History of Libraries in the Western World* (Metuchen, NJ: Scarecrow, 1984), 7.

12. Matthew Battles, *Library: An Unquiet History* (New York: Norton, 2003), 7.

13. A. C. Moorhouse, *Writing and the Alphabet* (London: Cobbett, 1946), 68–74.

14. Wiegand and Davis, *Encyclopedia of Library History*, 27.

15. Battles, *Library*, 12.

16. Foster Stockwell, *A History of Information Storage and Retrieval* (Jefferson, NC: McFarland, 2000), 11.

17. Robert Lamberton, *Hesiod* (New Haven, CT: Yale University Press, 1988), 64.

18. Lamberton, *Hesiod*, 39.

19. This phrase was coined by librarian and classicist H. Curtis Wright.

20. Lamberton, *Hesiod*, 41.

21. Merlin Donald, *Origins of the Modern Mind: Three Stages in the Evolution of Culture and Cognition* (Cambridge, MA: Harvard University Press, 1993), 275.

22. Plato, *Phaedrus*, trans. Benjamin Jowett (Champaign, IL: Project Gutenberg, 1999), http://www.gutenberg.org/dirs/etext99/phdrs10.txt.

23. Donald, *Origins of the Modern Mind*, 273.

24. Karl Popper draws much the same distinction between scientific and "prescientific" thought.

25. Donald, *Origins of the Modern Mind*, 343.

26. Donald, *Origins of the Modern Mind*, 274.

27. Strabo, quoted in Casson, *Libraries in the Ancient World*, 29.

28. Aristotle, *The Categories*, trans. E. M. Edghill (Champaign, IL: Project Gutenberg, 2000), http://www.gutenberg.org/dirs/etext00/arist10.txt.

29. Aristotle is said to have taught his students while, literally, walking around.

30. Leslie W Dunlap, *Readings in Library History* (New York: Bowker, 1972), 20.

31. Amr, quoted in Battles, *Library*, 22–23.

32. Ausonius, trans. John Evelyn, quoted in Raymond Irwin, *The Origins of the English Library* (Westport, CT: Greenwood, 1981), 133.

33. Martial, quoted in Dunlap, *Readings in Library History*, 32.

34. Ammanius, quoted in Dunlap, *Readings in Library History*, 34.

5. ILLUMINATING THE DARK AGE

1. Christopher de Hamel, *A History of Illuminated Manuscripts* (London: Phaidon, 1994), 22.

2. Michael E. Hobart and Zachary Sayre Schiffman, *Information Ages: Literacy, Numeracy, and the Computer Revolution* (Baltimore: Johns Hopkins University Press, 1998), 91.

3. Thomas Cahill, *How the Irish Saved Civilization: The Untold Story of Ireland's Heroic Role from the Fall of Rome to the Rise of Medieval Europe* (New York: Doubleday, 1995), 163.

4. De Hamel, *A History of Illuminated Manuscripts*, 11.

5. Bede, quoted in De Hamel, *A History of Illuminated Manuscripts*, 40.

6. Girardus Cambrensis, quoted (translated) by William J. Diebold, *Word and Image: An Introduction to Early Medieval Art* (Boulder, CO: Westview, 2000).

7. Cassiodorus, quoted in Leslie W. Dunlap, *Readings in Library History* (New York: Bowker, 1972), 68.

8. Matthew Battles, *Library: An Unquiet History* (New York: Norton, 2003), 60.

9. James J. O'Donnell, *Cassiodorus* (Berkeley: University of California Press, 1979).

10. Ivan Illich, *In the Vineyard of the Text: A Commentary to Hugh's Didascalicon* (Chicago: University of Chicago Press, 1996), 54.

11. Walter J. Ong, *Rhetoric, Romance, and Technology: Studies in the Interaction of Expression and Culture* (Ithaca, NY: Cornell University Press, 2012), 119.

12. Saint Benedict, quoted in Karl Christ, *The Handbook of Medieval Library History* (Metuchen, NJ: Scarecrow, 1984), 17.

13. Charlemagne, "De Litteris Colendis" [Letter to Baugaulf of Fulda], ca. 780–800.

14. Alcuin unwittingly anticipated use of the term *walled garden* by modern-day software developers—that is, as a sealed information environment (AOL, for instance, is a classic example of a networked "walled garden").

15. J. J. O'Connor and E. F. Robertson. "Alcuin of York," last updated November 1999, https://mathshistory.st-andrews.ac.uk/Biographies/Alcuin/.

16. Battles, *Library*, 79.

17. Ivan Illich and Barry Sanders. *ABC: The Alphabetization of the Popular Mind* (New York: Vintage Books, 1989), 40.

18. James Westfall Thompson. *The Medieval Library* (New York: Hafner, 1957), 350.

19. Violet Moller, *The Map of Knowledge: A Thousand-Year History of How Classical Ideas Were Lost and Found* (New York: Doubleday, 2019), 70.

20. Thompson, *The Medieval Library*, 348–50.

21. Thompson, *The Medieval Library*, 357.

22. Edward Gibbon, *The Decline and Fall of the Roman Empire* (New York: Modern Library, 2005), 1139.

23. Moller, *The Map of Knowledge*, 99.

24. Moller, *The Map of Knowledge*, 102.

25. Moller, *The Map of Knowledge*, 117.

26. Moller, *The Map of Knowledge*, 248.

6. A STEAM ENGINE OF THE MIND

1. Thomas Franklin Carter, *The Invention of Printing in China and Its Spread Westward*, 2nd ed. (New York: Ronald Press, 1955), 176–78.

2. Whit Andrews, personal correspondence with author, April 7, 2006.

3. Samuel A. Ives and Hellmut Lehman-Haupt, *An English 13th Century Bestiary, a New Discovery in the Technique of Medieval Illumination* (New York: H. P. Kraus, 1942).

4. Brian Stock, *The Implications of Literacy: Written Language and Models of Interpretation in the Eleventh and Twelfth Centuries* (Princeton, NJ: Princeton University Press, 1983), 32.

5. Stock, *The Implications of Literacy*, 18.

6. Stock, *The Implications of Literacy*, 62.

7. Émile Durkheim, *The Rules of the Sociological Method*, trans. W. D. Halls (New York: Free Press, 1982 [1895]), chap. 5.

8. Stock, *The Implications of Literacy*, 88.

9. Stock, *The Implications of Literacy*, 88.

10. Not to be confused with the modern stationer, the medieval stationer performed a complicated role involving taxation, quality assurance, and distribution of master copies of important texts.

11. Lucien Febvre and Henri-Jean Martin, *The Coming of the Book: The Impact of Printing, 1450–1800* (London: Verso, 1990), 167.

12. Johann Koelhoff, quoted in Febvre and Martin, *The Coming of the Book*, 171.

13. Indulgences were remittances granted by the church to allay punishment for sins. The practice of selling indulgences was eventually banned by Pope Pius VI in 1567.

14. Febvre and Martin, *The Coming of the Book*, 173.

15. Febvre and Martin, *The Coming of the Book*, 79.

16. Febvre and Martin, *The Coming of the Book*, 96.

17. Xylography was a popular printing technique using woodcut engravings.

18. Febvre and Martin, *The Coming of the Book*, 97.

19. John Henry Newman, *The Works of Cardinal Newman: The Idea of a University Defined and Illustrated* (London: Longmans, Green, 1917), 143.

20. Febvre and Martin, *The Coming of the Book*, 261.

21. Foster Stockwell, *A History of Information Storage and Retrieval* (Jefferson, NC: McFarland, 2008), 47.

22. George P. Landow, *Hypertext 2.0* (Baltimore: Johns Hopkins University Press, 1997), 21.

23. Elizabeth Eisenstein, *The Printing Press as an Agent of Change* (Cambridge: Cambridge University Press, 1980), 303.

24. Eisenstein, *The Printing Press*, 304.

25. Leonard Shlain, *The Alphabet versus the Goddess: The Conflict between Word and Image* (New York: Penguin, 1999), 325.

26. Shlain, *The Alphabet versus the Goddess*, 331.

27. John Lothrop Motley, *The Rise of the Dutch Republic, 1555–1566* (Champaign, IL: Project Gutenberg, 2006), https://www.gutenberg.org/files/4811/4811-h/4811-h.htm.

28. Shlain, *The Alphabet versus the Goddess*, 345.

29. Shlain, *The Alphabet versus the Goddess*, 372.

30. Shlain, *The Alphabet versus the Goddess*, 375.

7. THE ASTRAL POWER STATION

1. Frances Amelia Yates, *The Art of Memory* (London: Routledge, 1999), 137.

2. Rhodri Lewis, *From Athens to Elsinore: The Early Modern Art of Memory, Reconsidered* (Berlin: Max-Planck-Institut für Wissenschaftsgeschichte 2006), 5–7.

3. Aristotle, *De anima*, quoted in Yates, *The Art of Memory*, 32.

4. Thomas Aquinas, *Summa theologica* (New York: Benziger, 1922).

5. Willis, quoted in Lewis, *From Athens to Elsinore*, 19.

6. Yates, *The Art of Memory*, 121.

7. Yates, *The Art of Memory*, 80.

8. Yates, *The Art of Memory*, 113.

9. Yates, *The Art of Memory*, 123.

10. Yates, *The Art of Memory*, 172.

11. Lewis, *From Athens to Elsinore*, 10–12.

12. Yates, *The Art of Memory*, 158.

13. Yates, *The Art of Memory*, 127.

14. Bruno, quoted in Yates, *The Art of Memory*, 200.

15. Bruno, quoted in Yates, *The Art of Memory*, 228.

16. Yates, *The Art of Memory*, 223.

17. Like the Gnostics, Bruno believed that thirty was a mystical number (representing, for example, the number of disciples of John the Baptist and the number of eons). Thirty is also deeply associated with magical practices like those of Simon Magus.

18. Yates, *The Art of Memory*, 212–13.

19. Bruno, quoted in Yates, *The Art of Memory*, 217.

20. See chapter 1.

21. Lewis, *From Athens to Elsinore*, 14.

22. Aubrey, quoted in Yates, *The Art of Memory*, 370.

23. Bacon, *Advancement of Learning*, II, sv. 2; in *The Works of Francis Bacon* (Cambridge: Cambridge University Press, 2011), 398–99.

24. Francis Bacon, *Novum organum* (New York: P. F. Collier, 1902), 2.26.

25. Bacon, *Novum organum*.

26. Bacon, *Novum organum*.

27. The title is a play on Aristotle's *Organon*.

28. Bacon, *Novum organum*.

29. Bacon, *Novum organum*.

30. Bacon, preface to *New Atlantis and The Great Instauration* (Hoboken, NJ: Wiley-Blackwell, 2017).

31. Bacon, *Sylva sylvarum, or, a Natural History in Ten Centuries* (London: Bennet Griffin, 1683).

32. Bacon, preface to *New Atlantis and The Great Instauration*.

33. Lewis, *From Athens to Elsinore*, chap. 5.

34. Lewis, *From Athens to Elsinore*, 224.

35. Wilkins, quoted in Lewis, *From Athens to Elsinore*, 230.

36. Lewis, *From Athens to Elsinore*, 237.

37. Wilkins, quoted in Barbara J. Shapiro, *John Wilkins 1614–1672: An Intellectual Biography* (Berkeley: University of California Press, 2021), 49.

38. Jorge Luis Borges, "The Analytical Language of John Wilkins," trans. Douglas Crockford, https://www.crockford.com/wilkins.html.

8. THE ENCYCLOPEDIC REVOLUTION

1. Lucien Febvre and Henri-Jean Martin, *The Coming of the Book: The Impact of Printing, 1450–1800* (London: Verso, 1990), 262.

2. Thomas Heywood, *Gunaikeion*, as quoted in translation in Nonna Crook and Neil Rhodes, "The Daughters of Memory: Thomas Heywood's *Gunaikeion* and the Female Computer," in *The Renaissance Computer*, ed. Jonathan Sawday and Neil Rhodes (London: Routledge, 2000), 1317.

3. Heywood, *Gunaikeion*, quoted in Crook and Rhodes, "The Daughters of Memory," 137.

4. Heywood, *Gunaikeion*, quoted in Crook and Rhodes, "The Daughters of Memory," 136.

5. Crook and Rhodes, "The Daughters of Memory," 143.

6. Heywood, quoted in Crook and Rhodes, "The Daughters of Memory," 143.

7. Heywood, quoted in Crook and Rhodes, "The Daughters of Memory," 143.

8. Crook and Rhodes, "The Daughters of Memory," 145.

9. Stockwell, *A History of Information Storage and Retrieval*, 89–90

10. Brian Stock, *The Implication of Literacy: Written Language and Models of Interpretation in the Eleventh and Twelfth Centuries* (Princeton, NJ: Princeton University Press, 1983), 31.

9. THE MOOSE THAT ROARED

1. David G. Post, "Mr. Jefferson's Moose and the Law of Cyberspace," *FindLaw*, n.d., https://supreme.findlaw.com/legal-commentary/jeffersons-moose-and-the-law-of-cyberspace-part-i.html.

2. Harriet Ritvo, *The Platypus and the Mermaid, and Other Figments of the Classifying Imagination* (Cambridge, MA: Harvard University Press, 1997), 53.

3. Cf. Durkheim's work on primitive classification (see chapter 2).

4. Graham Shawcross, "Buffon's American Degeneracy," January 20, 2012. https://grahamshawcross.com/2012/01/20/buffons-american-degeneracy-theory/.

5. George Louis Le Clerc, *Histoire naturelle*, trans. William Smellie (London: W. Strahan and T. Cadell, 1785), 115.

6. Thomas Jefferson, "Letter to Dr. John Manners," February 22, 1814, National Archives, https://founders.archives.gov/documents/Jefferson/03-07-02-0132.

7. Jefferson, "Letter to Dr. John Manners."

10. THE INDUSTRIAL LIBRARY

1. Edmund Lester Pearson, as quoted in Matthew Battles, *Library: An Unquiet History* (New York: Norton, 2003), 14.

2. Battles, *Library*, 121–22.

3. Markus Krajewski, *Paper Machines: About Cards & Catalogs, 1548–1929.* Cambridge, MA: MIT Press, 2011.

4. George Dyson, *Darwin among the Machines: The Evolution of Global Intelligence* (New York: Basic Books, 2012).

5. Lewis Fagan, as quoted in Teresa Negrucci, "Historiography of Antonio Panizzi," 2001, https://pages.gseis.ucla.edu/faculty/maack/Documents/Panizzi.doc.

6. Elaine Svenonius, *The Intellectual Foundation of Information Organization* (Cambridge, MA: MIT Press, 2000), 3.

7. Shirley Hyatt, "Developments in Cataloging and Metadata," in *International Yearbook of Library and Information Management 2003–2004: Metadata Applications and Management*, ed. G. E. Gorman and Daniel G. Dorner (London: Facet Publishing, 2003), http://www.oclc.org/research/publications/archive/2003/hyatt.pdf.

8. Seymour Lubetzky, "The Vicissitudes of Ideology and Technology in Anglo-American Cataloging since Panizzi and a Prospective Reformation of the Catalog for the Next Century," in *Seymour Lubetzky: Writings on the Classical Art of Cataloging*, ed. Elaine Svenonius and Dorothy McGarry (Englewood, CO: Libraries Unlimited, 2001), 423.

9. Anthony Panizzi, as quoted in Battles, *Library*, 131.

10. Charles A. Cutter, "Editorial," *Library Journal*, February 1883, 23–24.

11. One of the first known card catalogs belonged to Edward Gibbon, who recorded the contents of his considerable private library on the backs of playing cards. French librarians developed the first institutional card catalogs in the 1700s. The Smithsonian librarian Charles Jewett also envisioned a large distributed catalog with cards that could be shared between libraries.

12. Colin Burke, *Information and Intrigue: From Index Cards to Dewey Decimals to Alger Hiss* (Cambridge, MA: MIT Press, 2014), 15.

13. Throughout his life, Dewey advocated for a simplified phonetic spelling scheme as a way to obtain more efficiency in written communications.

14. Melvil Dewey, "Decimal Classification Beginning," *Library Journal* 45, no. 15 (February 1920): 151–54.

15. Melvil Dewey, *Librarianship as a Profession for College-Bred Women* (Boston: Library Bureau, 1886).

16. Gretchen Keer, "The Stereotype Stereotype," *American Libraries*, October 30, 2015.

17. Melvil Dewey, "Classification and Subject Index for Cataloguing and Arranging the Books and Pamphlets of a Library" (1876).

18. Dewey, "Classification and Subject Index."

19. C. A. Cutter, "A Notation for Small Libraries," in *Papers and Proceedings of the Seventh General Meeting of the American Library Association* (Boston: Rockwell and Churchill, 1885).

11. INFORMATION AS "SCIENCE"

1. Rayward, W. Boyd. "The Case of Paul Otlet, Pioneer of Information Science, Internationalist, Visionary: Reflections on Biography," *Journal of Librarianship and Information Science* 23 (September 1991): 135–45.

2. Paul Otlet, *Traité de documentation: Le livre sur le livre* (Brussels: D. Van Keerberghen, 1934), 428.

3. Otlet, as quoted (translated) in W. Boyd Rayward, *The Universe of Information: The Work of Paul Otlet for Documentation and International Organization* (Moscow: VINITI for the International Federation for Documentation, 1975), 113.

4. Universal Decimal Classification Consortium, "About Universal Decimal (UDC)," http://www.udcc.org/about.htm.

5. King Albert I, as quoted in Olivia Andersen's diary, September 6, 1913.

6. Paul Otlet, *La fin de la guerre* (The Hague: Martinus Nijhoff, 1914).

7. Otlet, as quoted (translated) in Ron Day, "Paul Otlet's Book and the Writing of Social Space," *Journal of the American Society for Information Science* 48, no. 4 (April 1997): 310–17.

8. Otlet, *Monde: essai d'universalisme* (Brussels: Editiones Mundaneum, 1935), 391.

9. Eugene Garfield, "Librarian vs. Documentalist," *Special Libraries*, 1953, http://www.garfield.library.upenn.edu/papers/librarianvsdocumentalisty1953.html.

10. Thomas Hapke, "Wilhelm Ostwald, the 'Brücke' (Bridge), and Connections to Other Bibliographic Activities at the Beginning of the Twentieth Century," conference on the History and Heritage of Science Information Systems, Pittsburgh, PA, October 23–25, 1998, http://www.tu-harburg.de/b/hapke/ostwald/ostwald.htm.

11. Michael Buckland, *Emanuel Goldberg and His Knowledge Machine* (Westport, CT: Libraries Unlimited, 2006).

12. H. G. Wells, *World Brain* (London: Methuen, 1938), 23–35.

13. Rayward, W. Boyd. "H. G. Wells's Idea of a World Brain: A Critical Reassessment," *Journal of the American Society for Information Science* 50, no. 7 (May 15, 1999): 557–73.

14. Wells, *World Brain*, 74–77.

15. "Watson Davis 1896–1967," *Science News* 92, no. 2 (1967): 28–29.

16. Colin Burke, *Information and Intrigue: From Index Cards to Dewey Decimals to Alger Hiss* (Cambridge, MA: MIT Press, 2014).

17. Burke, *Information and Intrigue*, 40–41.

18. Burke, *Information and Intrigue*, 43–45.

19. Burke, *Information and Intrigue*, 65–91.

20. Burke, *Information and Intrigue*, 141–200.

21. Mary Niles Maack, "The Lady and the Antelope: Suzanne Briet's Contribution to the French Documentation Movement," *Library Trends* 52, no. 4 (2004): 719–47.

22. Maack, "The Lady and the Antelope."

23. Suzanne Briet, *What Is Documentation? English Translation of the Classic French Text*, trans. Ronald E. Day, Laurent Martinet, and Hermina G. B. Anghelescu (Lanham, MD: Scarecrow, 2006).

24. Briet, *What Is Documentation?*

25. Briet, *What Is Documentation?*

26. Day, "Paul Otlet's Book," 55.

27. Ethel Johnson, as quoted in Robert V. Williams, "The Documentation and Special Libraries Movements in the United States, 1910–1960," in *Historical Studies in Information Science*, ed. Michael Keeble Buckland and Trudi Bellardo Hahn (Medford, NJ: ASIS/Information Today, 1998), 174.

28. Josephson, quoted in Williams, "The Documentation and Special Libraries Movements," 173.

29. Marion, quoted in Williams, "The Documentation and Special Libraries Movements," 174.

30. Sergey Brin and Larry Page, "The Anatomy of a Large-Scale Hypertextual Web Search Engine," http://www-db.stanford.edu/~backrub/google.html.

31. Paul Kahn, personal correspondence with author, April 6, 2006.

12. THE WEB THAT WASN'T

1. Bush's prodigious résumé included the invention of an early analog computer, stints as a vice president at MIT, president of the Carnegie Institution, chairman of the National Defense Council, founder of the National Science Foundation, and cofounder of Raytheon. During World War II, he served as a close adviser to President Franklin D. Roosevelt and played an instrumental role in the Manhattan Project.

2. Ted Nelson, "As We Will Think," in *From Memex to Hypertext: Vannevar Bush and the Mind's Machine*, ed. James M. Nyce and Paul Kahn (San Diego: Academic Press, 1991), 148.

3. James M. Nyce and Paul Kahn, "A Machine for the Mind: Vannevar Bush's Memex," in *From Memex to Hypertext*, ed. James M. Nyce and Paul Kahn (San Diego: Academic Press, 1991), 45–46.

4. Vannevar Bush, "The Inscrutable Thirties," in *From Memex to Hypertext*, ed. James M. Nyce and Paul Kahn (San Diego: Academic Press, 1991), 67.

5. Vannevar Bush, "Memex Revisited," in *From Memex to Hypertext*, ed. James M. Nyce and Paul Kahn (San Diego: Academic Press, 1991), 201.

6. Michael K. Buckland, *Emanuel Goldberg and His Knowledge Machine: Information, Invention, and Political Forces* (Westport, CT: Libraries Unlimited, 2006).

7. Vannevar Bush, quoted in Nyce and Kahn, "A Machine for the Mind," 42.

8. Vannevar Bush, "As We May Think," *Atlantic Monthly*, July 1945, http://www.theatlantic.com/doc/194507/bush.

9. Michael Buckland has suggested that the now-famous display depicted in Alfred D. Cimi's illustration of the Memex in *Life* draws its inspiration from a real-life workstation already developed by Leonard G. Townsend in 1938.

10. Bush, "As We May Think."

11. Bush, "As We May Think."

12. Bush, "As We May Think."

13. Stigmergy is the process of altering external environments to introduce constraints on social behavior (see chapter 1).

14. Microsoft's MyLifeBits program and Texas A&M's Walden's Path project both attempt to apply Bush's vision of path-based navigation to digital environments.

15. Bush, quoted in Nyce and Kahn, "A Machine for the Mind," 61.

16. Nelson, "As We Will Think," 148.

17. Bush, "Memex Revisited," 201.

18. Bush, quoted in Nyce and Kahn, "A Machine for the Mind," 62.

19. Bush, quoted in Nyce and Kahn, "A Machine for the Mind," 135.

20. "Memex II" was originally slated to appear in the *Atlantic Monthly*, but a series of editorial disagreements and convoluted contractual disputes involving *Life* and *Fortune* ultimately precluded its publication. Bush himself felt that the essay was never quite complete. He later resurrected some of its themes in a subsequent essay, "Memex Revisited," published in his 1967 book *Science Is Not Enough*.

21. Herman H. Goldstine, *The Computer from Pascal to von Neumann* (Princeton, NJ: Princeton University Press, 1993), 119.

22. Claude Shannon, "A Mathematical Theory of Computation," *The Bell System Technical Journal* 27 (October 1948): 379–423, 623–56.

23. Francis Bello, "Information Theory," *Fortune*, December 1953, 136–58.

24. Licklider, J. C. R. "Memorandum for Members and Affiliates of the Intergalactic Computer Network." Kurzweil AI Net (republished December 11, 2011), April 23, 1963.

25. Engelbart, "Augmenting Human Intellect: A Conceptual Framework," October 1962, https://dougengelbart.org/content/view/138#9.

26. Engelbart, "Augmenting Human Intellect."

27. Engelbart, "Augmenting Human Intellect."

28. Engelbart, "Augmenting Human Intellect."

29. Theodor H. Nelson, *Literary Machines 93.1* (Sausalito, CA: Mindful Press, 1992), 0/3.

30. James Gillies and Robert Cailliau, *How the Web Was Born: The Story of the World Wide Web* (Oxford: Oxford University Press, 2000), 100.

31. Paul Kahn, personal correspondence with author, 2005.

32. Theodor H. Nelson, quoted in Gillies and Cailliau, *How the Web Was Born*, 101.

33. Theodor H. Nelson, speech at 1995 Vannevar Bush Symposium, quoted in Gillies and Cailliau, *How the Web Was Born*, 100.

34. Nelson, *Literary Machines 93.1*, 1/24.

35. Nelson, *Literary Machines 93.1*, 1/20.

36. Nelson, *Literary Machines 93.1*, 1/11.

37. Theodor H. Nelson, *Computer Lib: Dream Machines* (Redmond, WA: Tempus Books, 1987).

38. Nelson, *Literary Machines 93.1*, 0/6.

39. Steven Carmody et al. "A Hypertext Editing System for the /360," in *Pertinent Concepts in Computer Graphics: Proceedings of the Second University of Illinois Conference on Computer Graphics*, ed. Michael Faiman and Jürg Nievergelt (Champaign: University of Illinois Press, 1969), 291–330.

40. Steven J. DeRose and Andries van Dam, "Document Structure and Markup in the FRESS Hypertext System," *Markup Languages* 1, no. 1 (1999): 7–32.

41. James V. Catano, "Poetry and Computers: Experimenting with the Communal Text," *Computers and the Humanities* 13 (1979): 269–75.

42. Nicole Yankelovich, personal correspondence with the author, April 7, 2006.

43. George P. Landow, personal correspondence with the author, March 19, 2006.

44. George P. Landow, *Hypertext 2.0* (Baltimore: Johns Hopkins University Press, 1997), 55.

45. Landow, *Hypertext 2.0*, 2.

46. Landow, *Hypertext 2.0*, 73.

47. Landow, *Hypertext 2.0*, 85.

48. Gregory Ulmer, quoted in Landow, *Hypertext 2.0*, 59.

49. Landow, personal correspondence with the author, March 19, 2006.

50. Landow, personal correspondence with the author, March 19, 2006.

51. Wendy Hall, quoted in Gillies and Cailliau. *How the Web Was Born,* 128.

52. *Enquire within upon Everything*, quoted in Gillies and Cailliau, *How the Web Was Born*, 169.

53. Internet Live Stats, Internetlivestats.com, accessed January 31, 2022.

54. Theodor H. Nelson, "I Don't Buy In," http://ted.hyperland.com/buyin.txt.

55. MacKenzie Wark, quoted in Ben Vershbow, "Small Steps toward an N-Dimensional Reading/Writing Space," Institute for the Future of the Book, December 6, 2006, http://www.futureofthebook.org/blog/archives/2006/12/small_steps_toward_an_n-dimensional.html.

13. MEMORIES OF THE FUTURE

1. "One Laptop per Child," n.d., http://laptop.org/.

2. Walter J. Ong, *Rhetoric, Romance, and Technology: Studies in the Interaction of Expression and Culture* (Ithaca, NY: Cornell University Press, 2012).

3. Melvin Kranzberg, "Technology and History: Kranzberg's Laws," *Technology and Culture* 27, no. 3 (1986): 547.

4. Steven Pinker, *The Language Instinct: How the Mind Creates Language* (New York: HarperPerennial, 2010).

5. "2006 Literacy," UNESCO Education, 2006, https://en.unesco.org/gem-report/2006-literacy.

6. Thomas Vanderwal, "Folksonomy Coinage and Definition," Vanderwal.Net, February 2, 2007, https://vanderwal.net/folksonomy.html.

7. Davis S. Alberts and Richard E. Hayes, *Power to the Edge: Command, Control in the Information Age* (Washington, DC: CCRP, 2003).

8. John Seely Brown and Paul Duguid, *The Social Life of Information* (Boston: Harvard Business School Press, 2000), 170.

9. Kevin Kelly, *Out of Control: The New Biology of Machines, Social Systems, and the Economic World* (Reading, MA: Perseus Books, 1994), 45.

10. Francis Fukuyama, *The Great Disruption: Human Nature and the Reconstitution of Social Order* (New York: Touchstone, 2000).

11. John Markoff, *What the Doormouse Said: How the Sixties Counterculture Shaped the Personal Computer Industry* (New York: Penguin, 2006).

Bibliography

Alberts, David S., and Richard E. Hayes. *Power to the Edge: Command and Control in the Information Age.* Washington, DC: CCRP, 2003.

Allman, William F. *The Stone Age Present: How Evolution Has Shaped Modern Life: From Sex, Violence, and Language to Emotions, Morals, and Communities.* New York: Simon & Schuster, 1994.Anderson, Olivia. "Diary." n.d. In Hendrik Christian Anderson papers, 1844–1940. Washington, D.C.: Library of Congress, Manuscript Division.

Aquinas, Thomas. *Summa theologica.* New York: Benziger, 1922.

Aristotle. *The Categories.* Translated by E. M. Edghill. Champaign, IL: Project Gutenberg, 2000. http://www.gutenberg.org/dirs/etext00/arist10.txt.

Bacon, Francis. *New Atlantis and The Great Instauration.* Edited by Jerry Weinberger. Hoboken, NJ: Wiley-Blackwell, 2017.

——. *Novum organum.* New York: P. F. Collier, 1902. https://openlibrary.org/books/OL7059000M/Novum_Organum.

——. *Sylva sylvarum, or, a Natural History in Ten Centuries.* London: Bennet Griffin, 1683. https://www.loc.gov/item/95202443/.

——. *The Works of Francis Bacon.* Edited by James Spedding, Robert Leslie Ellis, and Douglas Denon Heath. Cambridge: Cambridge University Press, 2011. https://doi.org/10.1017/CBO9781139149570.

Battles, Matthew. *Library: An Unquiet History.* New York: Norton, 2003.

Baudin, Louis. *A Socialist Empire: The Incas of Peru.* Princeton, NJ: Van Nostrand, 1961.

Bello, Francisco. "Information Theory." *Fortune,* December 1953, 136–58.

Berlin, Brent. *Ethnobiological Classification: Principles of Categorization of Plants and Animals in Traditional Societies.* Princeton, NJ: Princeton University Press, 1992.

Birkerts, Sven. *The Gutenberg Elegies: The Fate of Reading in an Electronic Age.* New York: Ballantine, 1995.

Bloom, Howard K. *Global Brain: The Evolution of Mass Mind from the Big Bang to the 21st Century.* New York: Wiley, 2001.

Borges, Jorge Luis. "The Analytical Language of John Wilkins." Translated by Douglas Crockford. March 8, 2019. https://www.crockford.com/wilkins.html.

Briet, Suzanne. *What Is Documentation? English Translation of the Classic French Text.* Translated by Ronald E. Day, Laurent Martinet, and Hermina G. B. Anghelescu. Lanham, MD: Scarecrow, 2006.

Brin, Sergey, and Larry Page. "The Anatomy of a Large-Scale Hypertextual Web Search Engine." N.d. http://www-db.stanford.edu/~backrub/google.html.

Brown, John Seely, and Paul Duguid. *The Social Life of Information.* Boston: Harvard Business School Press, 2000.

Brown, Roger. "How Shall a Thing Be Called?" *Psychology Review* 65 (1958): 14–21.

Buckland, Michael K. *Emanuel Goldberg and His Knowledge Machine: Information, Invention, and Political Forces.* Westport, CT: Libraries Unlimited, 2006.

Burke, Colin B. *Information and Intrigue: From Index Cards to Dewey Decimals to Alger Hiss.* Cambridge, MA: MIT Press, 2014.

Bush, Vannevar. "As We May Think." *Atlantic Monthly*, July 1945. http://www.theatlantic
.com/doc/194507/bush.

——. "Memex Revisited." In *From Memex to Hypertext: Vannevar Bush and the Mind's
Machine*, edited by James M. Nyce and Paul Kahn, 197–216. San Diego: Academic
Press, 1991.

——. "The Inscrutable Thirties." In *From Memex to Hypertext: Vannevar Bush and the
Mind's Machine*, edited by James M. Nyce and Paul Kahn, 67–79. San Diego: Academic Press, 1991.

Cahill, Thomas. *How the Irish Saved Civilization: The Untold Story of Ireland's Heroic
Role from the Fall of Rome to the Rise of Medieval Europe*. New York: Doubleday,
1995.

——. *The Gifts of the Jews: How a Tribe of Desert Nomads Changed the Way Everyone
Thinks and Feels*. New York: Nan A. Talese, 1998.

Carmody, Steven, Walter Gross, Theodor H. Nelson, David Rice, and Andries van Dam.
"A Hypertext Editing System for the /360." In *Pertinent Concepts in Computer
Graphics: Proceedings of the Second University of Illinois Conference on Computer
Graphics*, edited by Michael Faiman and Jürg Nievergelt, 291–330. Champaign:
University of Illinois Press, 1969.

Carter, Thomas Franklin. *The Invention of Printing in China and Its Spread Westward*.
New York: Ronald Press, 1955.

Casson, Lionel. *Libraries in the Ancient World*. New Haven, CT: Yale University Press,
2001.

Castells, Manuel. *The Rise of the Network Society*. 2nd ed. Malden, MA: Wiley-Blackwell,
2009.

Catano, James V. "Poetry and Computers: Experimenting with the Communal Text."
Computers and the Humanities 13 (1979): 269–75.

Childe, Vere Gordon. *Man Makes Himself*. New York: New American Library, 1951.

Christ, Karl. *The Handbook of Medieval Library History*. Metuchen, NJ: Scarecrow, 1984.

Christakis, Nicholas A., and James H. Fowler. *Connected: The Surprising Power of Our
Social Networks and How They Shape Our Lives, How Your Friends' Friends' Friends
Affect Everything You Feel, Think, and Do*. New York: Little, Brown, 2011.

Clark, Andy. *Being There: Putting Brain, Body, and World Together Again*. Cambridge,
MA: MIT Press, 2001.

Clark, John Willis. *On the Vatican Library of Sixtus IV*. Cambridge: Cambridge Anti-
quarian Society, 1899.

Crook, Nonna, and Neil Rhodes. "The Daughters of Memory: Thomas Heywood's *Gunaikeion* and the Female Computer." In *The Renaissance Computer*, edited by Jonathan
Sawday and Neil Rhodes, 135–47. London: Routledge, 2000.

Cutter, Charles A. "A Notation for Small Libraries." In *Papers and Proceedings of the Seventh General Meeting of the American Library Association*, 14–16. Boston: Rockwell and Churchill, 1885.

——. "Editorial." *Library Journal*, February 1883, 23–24.

Day, Ron. "Paul Otlet's Book and the Writing of Social Space." *Journal of the American
Society for Information Science* 48, no. 4 (April 1997): 310–17.

De Hamel, Christopher. *A History of Illuminated Manuscripts*. London: Phaidon, 1984.

DeRose, Steven J., and Andries van Dam. "Document Structure and Markup in the
FRESS Hypertext System." *Markup Languages* 1, no. 1 (1999): 7–32.

Dewey, Melvil. "Classification and Subject Index for Cataloguing and Arranging the
Books and Pamphlets of a Library." Amherst, MA, 1876.

——. "Decimal Classification Beginning." *Library Journal* 45, no. 15 (February 1920):
151–54.

——. *Librarianship as a Profession for College-Bred Women.* Boston: Library Bureau, 1886.

Diebold, William J. *Word and Image: An Introduction to Early Medieval Art.* Boulder, CO: Westview, 2000.

Donald, Merlin. *Origins of the Modern Mind: Three Stages in the Evolution of Culture and Cognition.* Cambridge, MA: Harvard University Press, 1993.

Dunlap, Leslie W. *Readings in Library History.* New York: Bowker, 1972.

Durkheim, Émile, and Steven Lukes. *The Rules of Sociological Method: And Selected Texts on Sociology and Its Method.* Translated by W. D. Halls. New York: Free Press, 1982 [1895]).

Durkheim, Émile, and Marcel Mauss. *Primitive Classification.* Chicago: University of Chicago Press, 1963.

Dyson, George. *Darwin among the Machines: The Evolution of Global Intelligence.* New York: Basic Books, 2012.

Eisenstein, Elizabeth L. *The Printing Press as an Agent of Change: Communications and Cultural Transformations in Early-Modern Europe, Volumes I and II [Complete in One Volume].* Cambridge: Cambridge University Press, 2009.

Engelbart, Douglas. "Augmenting Human Intellect: A Conceptual Framework." October 1962. https://dougengelbart.org/content/view/138#9.

Febvre, Lucien, and Henri-Jean Martin. *The Coming of the Book: The Impact of Printing, 1450–1800.* London: Verso, 1990.

Ferguson, Niall. *The Square and the Tower: Networks and Power, from the Freemasons to Facebook.* New York: Penguin, 2018.

Fukuyama, Francis. *The Great Disruption: Human Nature and the Reconstitution of Social Order.* New York: Touchstone, 2000.

Gamble, Clive. "Palaeolithic Society and the Release from Proximity," *World Archaeology* 29, no. 3 (1998): 426–49.

Garfield, Eugene. "Librarian versus Documentalist." *Special Libraries,* 1953. http://www.garfield.library.upenn.edu/papers/librarianvsdocumentalisty1953.html.

Gibbon, Edward. *The Decline and Fall of the Roman Empire.* New York: Modern Library, 2005.

Gillies, James, and Robert Cailliau. *How the Web Was Born: The Story of the World Wide Web.* Oxford: Oxford University Press, 2000.

Gladwell, Malcolm. *The Tipping Point: How Little Things Can Make a Big Difference.* Boston: Back Bay Books, 2002.

Goldstine, Herman Heine. *The Computer from Pascal to von Neumann.* Princeton, NJ: Princeton University Press, 1993.

Gould, Stephen Jay. "Bacon, Brought Home." *Natural History* 108 (1999): 28–32, 72–78.

——. Foreword to *Five Kingdoms: An Illustrated Guide to the Phyla of Life on Earth,* 3rd ed., by Lynn Margulis and Karlene V. Schwartz. New York: Freeman, 1998.

Greenspan, Stanley I., and Stuart G. Shanker. *The First Idea: How Symbols, Language, and Intelligence Evolved from Our Primate Ancestors to Modern Humans.* Lebanon, IN: Da Capo, 2006.

Guthrie, R. Dale. *The Nature of Paleolithic Art.* Chicago: University of Chicago Press, 2005.

Hapke, Thomas. "Wilhelm Ostwald, the 'Brücke' (Bridge), and Connections to Other Bibliographic Activities at the Beginning of the Twentieth Century." Conference on the History and Heritage of Science Information Systems, Pittsburgh, PA, October 23–25, 1998. http://www.tu-harburg.de/b/hapke/ostwald/ostwald.htm.

Harris, Michael Hope. *History of Libraries in the Western World.* Metuchen, NJ: Scarecrow, 1984.

Hawkins, Jeff, and Sandra Blakeslee. *On Intelligence.* New York: Owl Books, 2005.

Henry, John. *Knowledge Is Power: How Magic, the Government and an Apocalyptic Vision Helped Francis Bacon to Create Modern Science.* London: Icon Books, 2017.

Hillis, Danny. "The Big Picture." *Wired* 6, no. 1 (1998). http://www.wired.com/wired/archive/6.01/hillis.html.

Hobart, Michael E., and Zachary S. Schiffman. *Information Ages: Literacy, Numeracy, and the Computer Revolution.* Baltimore: Johns Hopkins University Press, 1998.

Hyatt, Shirley. "Developments in Cataloging and Metadata." In *International Yearbook of Library and Information Management 2003–2004: Metadata Applications and Management,* edited by G. E. Gorman and Daniel G. Dorner. Lanham, MD: Scarecrow, 2004. https://www.oclc.org/content/dam/research/publications/library/2003/hyatt.pdf.

Illich, Ivan. *In the Vineyard of the Text: A Commentary to Hugh's Didascalicon.* Chicago: University of Chicago Press, 1996.

Illich, Ivan, and Barry Sanders. *ABC: The Alphabetization of the Popular Mind.* New York: Vintage Books, 1989.

Irwin, Raymond. *The Origins of the English Library.* Westport, CT: Greenwood, 1981.

Ives, Samuel A., and Hellmut Lehman-Haupt. *An English 13th Century Bestiary, a New Discovery in the Technique of Medieval Illumination.* New York: H. P. Kraus, 1942.

Jefferson, Thomas. "Letter to Dr. John Manners." February 22, 1814. National Archives. https://founders.archives.gov/documents/Jefferson/03-07-02-0132.

Johnson, Steven. *Interface Culture: How New Technology Transforms the Way We Create and Communicate.* San Francisco: HarperEdge, 1997.

Keer, Gretchen. "The Stereotype Stereotype." *American Libraries,* October 30, 2015.

Kelly, Kevin. *Out of Control: The New Biology of Machines, Social Systems, and the Economic World.* Reading, MA: Perseus Books, 1994.

Krajewski, Markus. *Paper Machines: About Cards & Catalogs, 1548–1929.* Cambridge, MA: MIT Press, 2011.

Kranzberg, Melvin. "Technology and History: Kranzberg's Laws." *Technology and Culture* 27, no. 3 (1986): 544–60.

Kuhn, Steven L., Mary C. Stiner, David S. Reese, and Erksin Güleç. "Ornaments of the Earliest Upper Paleolithic." In *Proceedings of the National Academy of Sciences* 98, no. 13 (2001): 7641–46.

Kurzweil, Ray. "The Law of Accelerating Returns." March 7, 2001. https://www.kurzweilai.net/the-law-of-accelerating-returns

Lakoff, George. *Women, Fire, and Dangerous Things: What Categories Reveal about the Mind.* Chicago: University of Chicago Press, 1987.

Lamberton, Robert. *Hesiod.* New Haven, CT: Yale University Press, 1988.

Landow, George P. *Hypertext 2.0.* Baltimore: Johns Hopkins University Press, 1997.

Le Clerc, George Louis. *Histoire naturelle.* Translated by William Smellie. London: W. Strahan and T. Cadell, 1785.

Levine, Rick, ed. *The Cluetrain Manifesto.* 10th anniversary ed. New York: Basic Books, 2009.

Lewis, Rhodri. *From Athens to Elsinore: The Early Modern Art of Memory, Reconsidered.* Berlin: Max-Planck-Institut für Wissenschaftsgeschichte, 2006.

Licklider, J. C. R. "Memorandum for Members and Affiliates of the Intergalactic Computer Network." Kurzweil AI Net (republished December 11, 2011), April 23, 1963.

Lovink, Geert. "The Archaeology of Computer Assemblage." *Mediamatic* 7, no. 1 (January 1, 1992). https://www.mediamatic.net/en/page/9116/the-archeology-of-computer-assemblage.

Lubetzky, Seymour in collaboration with Elaine Svenonius, "The Vicissitudes of Ideology and Technology in Anglo-American Cataloging Since Panizzi and a Prospective Reformation of the Catalog for the Next Century." In *The Future of Cataloging: In-*

sights from the Lubetzky Symposium, edited by Tschera Harkness Connell and Robert L. Maxwell, 3–11. Chicago: ALA, 2000); reprinted in *Seymour Lubetzky: Writings on the Classical Art of Cataloging*, edited by Elaine Svenonius and Dorothy McGarry, 419–30. Englewood, CO.: Libraries Unlimited, 2001.

Lyman, Peter, and Hal Varian. "How Much Information?" 2003. http://www.sims.berkeley.edu/how-much-info-2003.

Maack, Mary Niles. "The Lady and the Antelope: Suzanne Briet's Contribution to the French Documentation Movement." *Library Trends* 52, no. 4 (2004): 719–47.

Margulis, Lynn, and Karlene V. Schwartz. *Five Kingdoms: An Illustrated Guide to the Phyla of Life on Earth*. 3rd ed. New York: Freeman, 1998.

Markoff, John. *What the Dormouse Said: How the Sixties Counterculture Shaped the Personal Computer Industry*. New York: Penguin, 2006.

McCrone, John. *Going Inside: A Tour around a Single Moment of Consciousness*. London: Faber, 2000.

Moller, Violet. *The Map of Knowledge: A Thousand-Year History of How Classical Ideas Were Lost and Found*. New York: Doubleday, 2019.

Moorhouse, A. C. *Writing and the Alphabet*. London: Cobbett, 1946.

Motley, John Lothrop. *The Rise of the Dutch Republic, 1555–1566*. Champaign, IL: Project Gutenberg, 2006. https://www.gutenberg.org/files/4811/4811-h/4811-h.htm.

Negrucci, Teresa. "Historiography of Antonio Panizzi." 2001. https://www.gseis.ucla.edu/faculty/maack/Documents/Panizzi.doc.

Nelson, Theodor H. "As We Will Think." In *From Memex to Hypertext: Vannevar Bush and the Mind's Machine*, edited by James M. Nyce and Paul Kahn, 245–60. San Diego: Academic Press, 1991.

——. *Computer Lib: Dream Machines*. Rev. ed. Redmond, WA: Tempus Books, 1987.

——. "I Don't Buy In." N.d. http://ted.hyperland.com/buyin.txt.

——. *Literary Machines*. Sausalito, CA: Mindful Press, 1992.

Newman, John Henry. *The Works of Cardinal Newman: The Idea of a University Defined and Illustrated*. London: Longmans, Green, 1917.

Nyce, James M., and Paul Kahn. "A Machine for the Mind: Vannevar Bush's Memex." In *From Memex to Hypertext: Vannevar Bush and the Mind's Machine*, edited by James M. Nyce and Paul Kahn, 39–66. San Diego: Academic Press, 1991.

O'Connor, J. J., and E. F. Robertson. "Alcuin of York." Last updated November 1999. https://mathshistory.st-andrews.ac.uk/Biographies/Alcuin/.

O'Donnell, James J. *Cassiodorus*. Berkeley: University of California Press, 1979.

"One Laptop per Child." N.d. www.laptop.org.

Ong, Walter J. *Orality and Literacy: The Technologizing of the Word*. London: Methuen, 1982.

——. *Rhetoric, Romance, and Technology: Studies in the Interaction of Expression and Culture*. Ithaca, NY: Cornell University Press, 2012.

Otlet, Paul. *La fin de la guerre*. The Hague: Martinus Nijhoff, 1914.

——. *Monde: essai d'universalisme*. Brussels: Editiones Mundaneum, 1935.

——. *Traité de documentation: Le livre sur le livre*. Brussels: Editiones Mundaneum–Palais Mondial, 1934.

Pinker, Steven. *The Language Instinct: How the Mind Creates Language*. New York: Harper Perennial, 2010.

Plato. *Phaedrus*. Translated by Benjamin Jowett. Champaign, IL: Project Gutenberg, 1999. http://www.gutenberg.org/dirs/etext99/phdrs10.txt.

Post, David G. "Mr. Jefferson's Moose and the Law of Cyberspace." *FindLaw*, n.d. https://supreme.findlaw.com/legal-commentary/jeffersons-moose-and-the-law-of-cyberspace-part-i.html.

Rayward, W. Boyd. "H. G. Wells's Idea of a World Brain: A Critical Reassessment." *Journal of the American Society for Information Science* 50, no. 7 (May 15, 1999): 557–73.

——. "The Case of Paul Otlet, Pioneer of Information Science, Internationalist, Visionary: Reflections on Biography." *Journal of Librarianship and Information Science* 23 (September 1991): 135–45.

——. *The Universe of Information: The Work of Paul Otlet for Documentation and International Organization*. Moscow: VINITI for the International Federation for Documentation, 1975. http://hdl.handle.net/2142/651.

Reinsel, David, John Gantz, and John Rydning. *The Digitization of the World: From Edge to Core*. Framingham, MA: IDC, November 2018. https://www.seagate.com/files/www-content/our-story/trends/files/idc-seagate-dataage-whitepaper.pdf.

Rhodes, Neil, and Jonathan Sawday, eds. *The Renaissance Computer: Knowledge Technology in the First Age of Print*. New York: Routledge, 2000.

Rincon, Paul. "'Earliest Writing' Found in China." BBC News, April 17, 2003. http://news.bbc.co.uk/2/hi/science/nature/2956925.stm.

Ritvo, Harriet. *The Platypus and the Mermaid, and Other Figments of the Classifying Imagination*. Cambridge, MA: Harvard University Press, 1997.

Sessions, William A. *Francis Bacon Revisited*. New York: Twayne, 1996.

Shannon, Claude. "A Mathematical Theory of Computation." *The Bell System Technical Journal* 27 (October 1948): 379–423, 623–56.

Shapiro, Barbara J. *John Wilkins 1614–1672: An Intellectual Biography*. Berkeley: University of California Press, 2021.

Shawcross, Graham. "Buffon's American Degeneracy," January 20, 2012. https://grahamshawcross.com/2012/01/20/buffons-american-degeneracy-theory/

Shlain, Leonard. *The Alphabet versus the Goddess: The Conflict between Word and Image*. New York: Penguin, 1999.

Standage, Tom. *Writing on the Wall: Social Media, the First 2,000 Years*. New York: Bloomsbury, 2013.

Stock, Brian. *The Implications of Literacy: Written Language and Models of Interpretation in the Eleventh and Twelfth Centuries*. Princeton, NJ: Princeton University Press, 1983.

Stockwell, Foster. *A History of Information Storage and Retrieval*. Jefferson, NC: McFarland, 2000.

Surowiecki, James. *The Wisdom of Crowds: Why the Many Are Smarter than the Few and How Collective Wisdom Shapes Business, Economies, Societies, and Nations*. New York: Doubleday, 2004.

Svenonius, Elaine. *The Intellectual Foundation of Information Organization*. Cambridge, MA: MIT Press, 2000.

Svenonius, Elaine, and Dorothy McGarry, eds. *Seymour Lubetzky: Writings on the Classical Art of Cataloging*. Englewood, CO: Libraries Unlimited, 2001.

Teilhard de Chardin, Pierre. *The Future of Man*. New York: Image Books/Doubleday, 2004.

Thompson, James Westfall. *The Medieval Library*. New York: Hafner, 1957.

"2006 Literacy." UNESCO Education, 2006. https://en.unesco.org/gem-report/2006-literacy.

Universal Decimal Classification Consortium. "About Universal Decimal Classification (UDC)." N.d. https://udcc.org/index.php/site/page?view=about.

Van Schaik, Carel. "Why Are Some Animals So Smart?" *Scientific American*, April 2006. https://www.scientificamerican.com/article/why-are-some-animals-so-smart/.

Vanderwal, Thomas. "Folksonomy Coinage and Definition." Vanderwal.Net, February 2, 2007. https://vanderwal.net/folksonomy.html.

Vershbow, Ben. "Small Steps toward an N-Dimensional Reading/Writing Space." Institute for the Future of the Book, December 6, 2006. http://futureofthebook.org/blog/2006/12/06/small_steps_toward_an_n-dimensional/.

"Watson Davis 1896–1967." *Science News* 92, no. 2 (1967): 28–29.

Wells, H. G. *World Brain*. London: Methuen, 1938.

Wiegand, Wayne A., and Donald G. Davis, eds. *Encyclopedia of Library History*. New York: Garland, 1994.

Williams, Robert V. "The Documentation and Special Libraries Movements in the United States, 1910–1960." In *Historical Studies in Information Science*, edited by Michael Keeble Buckland and Trudi Bellardo Hahn, 173–79. Medford, NJ: ASIS/Information Today, 1998.

Wilson, Edward O. *Consilience: The Unity of Knowledge*. New York: Vintage, 1999.

Wolfe, Tom. "Digibabble, Fairy Dust and the Human Anthill." *Forbes ASAP* 164, no. 8 (1999): 213–27.

——. *The Electric Kool-Aid Acid Test*. New York: Picador, 2008.

Wright, Alex. *Cataloging the World: Paul Otlet and the Birth of the Information Age*. New York: Oxford University Press, 2014.

Wright, Robert. *Nonzero: The Logic of Human Destiny*. New York: Vintage, 2001.

Wurman, Richard Saul. *Information Architects*. Graphis, 1997.

Yates, Frances A. *The Art of Memory*. London: Routledge, 1999.

Index

9 781501 768675